Apollo 9

The NASA Mission Reports

Compiled from the NASA archives & Edited
by Robert Godwin

All rights reserved under article two of the Berne Copyright Convention (1971).
We acknowledge the financial support of the Government of Canada through the
Book Publishing Industry Development Program for our publishing activities.
Published by Apogee Books an imprint of Collector's Guide Publishing Inc., Box 62034, Burlington, Ontario, Canada, L7R 4K2
Printed and bound in Canada by Webcom Ltd of Toronto
Apollo 9 - The NASA Mission Reports
Edited by Robert Godwin
ISBN 1-896522-51-3
All photos courtesy of NASA

Introduction

Undoubtedly Apollo 8's resounding success was a hard act to follow. Given the fact that Borman Lovell and Anders didn't realise until the eleventh hour that they would be sent to the moon, it is hardly surprising that the crew of Apollo 9 probably didn't have much time to ponder the intangible and unpredictable fickle behaviour of the world media. While Apollo 8 was giving the world's population a new perspective, Colonel James McDivitt, Colonel David Scott and civilian Russell Schweickart were preparing to fly the world's first true spacecraft.

Built by the Grumman Corporation in Bethpage New York, the Lunar Module (LM) was built to perform only in the harsh environment of space. The LM had an astonishingly delicate appearance. It did not need to endure the immense heat of re-entry or abide by any of the traditional dictates of aerodynamics. It was a true space vehicle and Apollo 9 was to be the first time it would fly with a live crew.

One of the greatest ambitions of any test pilot is to be the first to fly a new vehicle, and just as Apollo 7 had been the first manned test of the Command and Service Modules (CSM) Apollo 9 would be the first real test of the LM, which had been appropriately christened "Spider" due to it's obvious insect-like appearance. Jim McDivitt and Rusty Schweickart would be the first people to ever fly autonomously in a spacecraft that couldn't bring them home if something went wrong. As Colonel McDivitt pointed out with wry humour, both ships could land — however only the CSM could land safely.

As might be expected in the 1960's, NASA was an organisation which had not yet become noted for waste and every single second of the flight of Apollo 9 was mapped out in precise detail, in fact it was a schedule that would intimidate the hardiest of test pilots. Every single system of the moon lander — environment, propulsion, navigation, communication — all needed to be tested and operated under as many conceivable situations as time allowed, and most importantly the flight required that the LM be flown detached from the CSM and then successfully re-docked. The crew of Apollo 9 had to prove that the two ships could fly miles apart and then relocate one another. All of these things had never been tried before and as is often the case in space, they involved potentially hazardous procedures. Assuming that the LM could be run through its paces successfully, McDivitt's crew were then expected to spend a couple of days conducting a photographic study of various targets on the ground.

Further on in this book you will find a copy of the time-line to which the crew had to adhere. It incorporates a staggering amount of work for ten days, but the crew of Apollo 9 accomplished it with a near perfect flight. They successfully conducted an EVA (or space-walk) which established that the crew could get back to the Command module if the docking mechanisms failed. They took an amazing stream of pictures of the docked spacecraft in orbit as well as some of the most beautiful and unique pictures ever taken of the LM as it floated above the blue oceans and white clouds of the Earth. The LM proved it was able to live up to the task of finding its way back and forth from the mothership and to cap it all off Apollo 9 provided the scientists back home an abundance of photographic data during the SO65 experiment that would be helping farmers, geologists, biologists and oceanographers for years to come.

In many histories of the Apollo program the flights of Apollo 9 and 7 often seem to be omitted. Certainly the drama of going to the moon was greatly responsible for the apparent interest in the other flights of Apollo, but without the test flights — without Apollo 9 — none of those other accomplishments would have been possible. The reader may notice some duplication of data in the following pages; this was a conscious decision on my part to retain the integrity of the original documents. Where the Press Kit may provide a rudimentary description of the hardware or flight plan, the Mission Reports are more in-depth. You may also notice that there are many new facts which were not included in the previous Apollo 8 book. This is partly due to the fact that Apollo 9 was test flying the LM, but also because an entirely different set of parameters had to be mapped, detailed and finally accomplished by the NASA team, to take this next, but critically important step to getting men to the moon.

Robert Godwin
(Editor)

The Command Module for Apollo 9 with some
of the people who manufactured it.

Special thanks to Stewart Bailey

Apollo 9
The NASA Mission Reports

(from the archives of the National Aeronautics and Space Administration)

For Jim, Dave and Rusty

Contents

APOLLO 9 PRESS KIT

MISSION OPERATION REPORT

LIST OF FIGURES

LIST OF TABLES

MISSION OPERATION REPORT (APOLLO 9) SUPPLEMENT

LIST OF FIGURES

POST LAUNCH MISSION OPERATION REPORT

LIST OF TABLES

PRESS KIT

FOR RELEASE: SUNDAY
February 23, 1969

RELEASE NO: 69-29

PROJECT: APOLLO 9

APOLLO 9 CARRIES LUNAR MODULE

Apollo 9, scheduled for launch at 11 a.m. EST, Feb. 28, from the National Aeronautics and Space Administration's Kennedy Space Center Launch Complex 39A, is the first manned flight of the Apollo spacecraft lunar module (LM).

The Earth-orbital mission will include extensive performance tests of the lunar module, a rendezvous of the lunar module with the command and service modules, and two hours of extravehicular activity by the lunar module pilot.

To the maximum extent possible, the rendezvous in Earth orbit will resemble the type of rendezvous that will take place in lunar orbit following a lunar landing. Rendezvous and docking of the lunar module with the command and service modules, extensive testing of the lunar module engines and other systems, and extravehicular activity are among the mission's objectives.

Apollo 9 crewmen are Spacecraft Commander James A. McDivitt, Command Module Pilot David R. Scott and Lunar Module Pilot Russell Schweickart. The mission will be the second space flight for McDivitt (Gemini 4) and Scott (Gemini 8) and the first for Schweickart.

Backup crew is comprised of Spacecraft Commander Charles Conrad, Jr., Command Module Pilot Richard F. Gordon and Lunar Module Pilot Alan L. Bean.

Mission events have been arranged in a work-day basis in what is perhaps the most ambitious NASA manned space mission to date. The first five days are packed with lunar module engine tests and systems checkouts, burns of the service module's 20,500-pound-thrust engine while the command/service module and lunar module are docked, the lunar module pilot's hand-over-hand transfer in space from the lunar module to the command module and back again, and rendezvous. The remainder of the 10-day, open-ended mission will be at a more leisurely pace.

The first day's mission activities revolve around docking the command module to the lunar module still attached to the Saturn V launch vehicle S-IVB third stage. When docking is complete and the tunnel joint between the two spacecraft is rigid, the entire Apollo spacecraft will be spring-ejected from the S-IVB. Maneuvering more than 2,000 feet away from the S-IVB, the Apollo 9 crew will observe the first of two restarts of the S-IVB's J-2 engine -- the second of which will boost the stage into an Earth-escape trajectory and into solar orbit.

Other first-day mission activities include a docked service propulsion engine burn to improve orbital lifetime and to test the command/service module (CSM) digital autopilot (DAD) during service propulsion system (SPS) burns.

The digital autopilot will undergo additional stability tests during the second work day when the SPS engine is ignited three more times. Also, the three docked burns reduce CSM weight to enhance possible contingency rescue of the lunar module during rendezvous using the service module reaction control thrusters.

A thorough checkout of lunar module systems takes up most of the third work day when the spacecraft commander and lunar module pilot transfer through the docking tunnel to the LM and power it up.

Among LM tests will be an out-of-plane docked burn of the descent stage engine under control of the LM digital autopilot, with the last portion of the burn manually throttled by the spacecraft commander.

After LM power-down and crew transfer back to the command module, the fifth docked service propulsion engine burn will circularize the orbit at 133 nautical miles as a base orbit for the LM active rendezvous two days later.

APOLLO 9 Launch Day

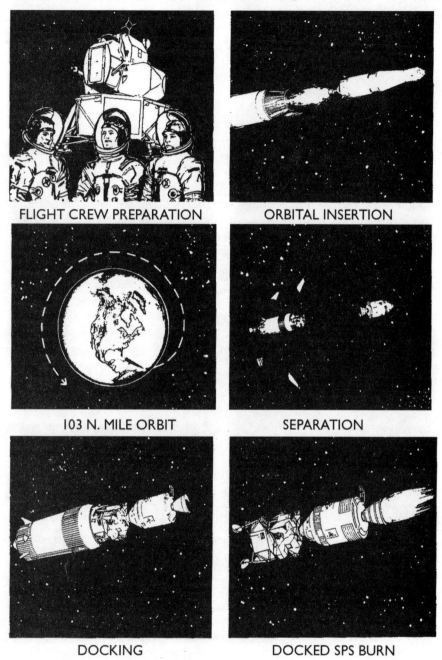

FLIGHT CREW PREPARATION

ORBITAL INSERTION

103 N. MILE ORBIT

SEPARATION

DOCKING

DOCKED SPS BURN

The fourth mission work day consists of further lunar module checks and extravehicular activity. The spacecraft commander and lunar module pilot again crawl through the tunnel to power the LM and prepare the LM pilot's extravehicular mobility unit (EMU -- EVA suit with life support backpack) for his two-hour stay outside the spacecraft.

Both spacecraft will be depressurized for the EVA and the LM pilot will climb out through the LM front hatch onto the "porch". From there, he will transfer hand-over-hand along handrails to the open command module side hatch and partly enter the cabin to demonstrate a contingency transfer capability.

The LM pilot will retrieve thermal samples on the spacecraft exterior and, returning to "golden slipper" foot restraints on the LM porch, will photograph both spacecraft from various angles and test the

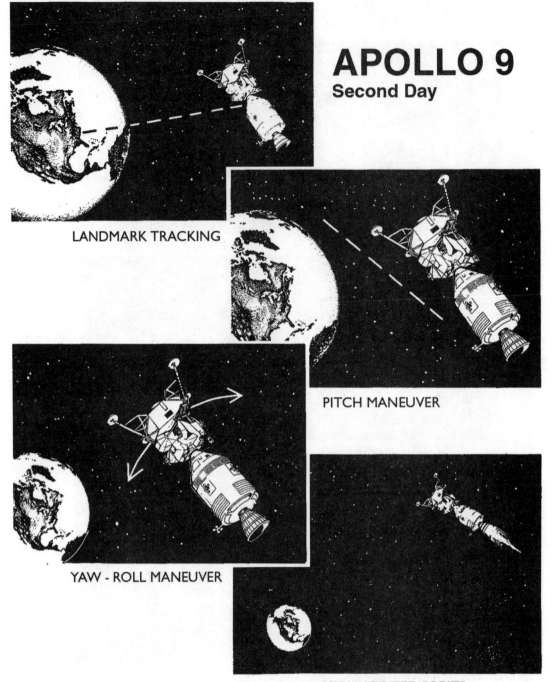

APOLLO 9
Second Day

LANDMARK TRACKING

PITCH MANEUVER

YAW - ROLL MANEUVER

HIGH APOGEE ORBITS

lunar surface television camera for about 10 minutes during a pass over the United States. This is a new model camera that has not been used in previous missions.

Both spacecraft will be closed and repressurized after the LM pilot gets into the LM, the LM will be powered down, and both crewmen will return to the command module.

The commander and LM pilot return to the LM the following day and begin preparations for undocking and a sequence simulating the checkout for a lunar landing descent.

A small thrust with the CSM reaction control system thrusters after separation from the LM places

APOLLO 9 Third Day

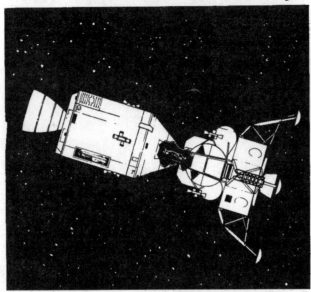

CREW TRANSFER

LM SYSTEM EVALUATION

the CSM in an orbit for a small-scale rendezvous, "mini-football", in which the maximum distance between the two spacecraft is about three nautical miles. The LM rendezvous radar is locked on to the CSM transponder for an initial test during this period.

One-half revolution after separation, the LM descent engine is ignited to place the LM in an orbit ranging 11.8 nautical miles above and below the CSM orbit, and after 1¼ revolutions it is fired a second time, to place the LM in an orbit 11 nautical miles above the CSM.

The LM descent stage will be jettisoned and the LM ascent stage reaction control thrusters will next be fired to lower LM perigee to 10 miles below the CSM orbit and set up conditions for circularization. Maximum LM-to-CSM range will be 95 nautical miles during this sequence.

APOLLO 9
Fourth Day

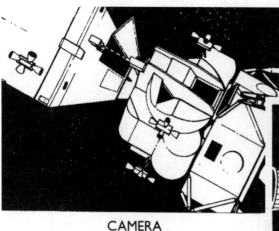

CAMERA

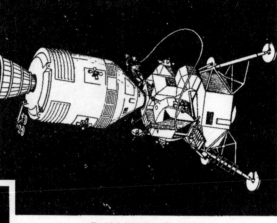

DAY-NIGHT EVA

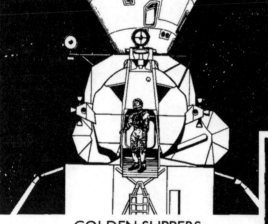

GOLDEN SLIPPERS

TV - TEXAS, FLORIDA

Circularization of the ascent stage orbit 10 nautical miles below the CSM and closing range will be the first duty in the mission for the 3,500-pound thrust ascent engine.

As the ascent stage approaches to some 20 nautical miles behind and 10 nautical miles below the CSM, the commander will thrust along the line of sight toward the CSM with the ascent stage RCS thrusters, making necessary midcourse corrections and braking maneuvers until the rendezvous is complete.

When the two vehicles dock, the ascent stage crew will prepare the ascent stage for a ground-commanded ascent engine burn to propellant depletion and transfer back into the command module. After the final undocking, the CSM will maneuver to a safe distance to observe the ascent engine depletion burn which will place the LM ascent stage in an orbit with an estimated apogee of 3,200 nautical miles.

APOLLO 9 Fifth Day

VEHICLES UNDOCKED

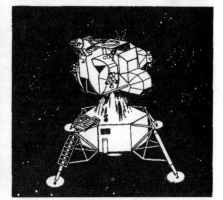

LM - BURNS FOR RENDEZVOUS

MAXIMUM SEPARATION

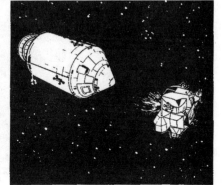

APS BURN

**FORMATION FLYING
& DOCKING**

**LM JETTISON
ASCENT BURN**

The sixth mission day is at a more leisurely pace, with the major event being a burn of the service propulsion system to lower perigee to 95 nautical miles for improved RCS thruster deorbit capability.

The seventh SPS burn is scheduled for the eighth day to extend orbital lifetime and enhance RCS deorbit capability by raising apogee to 210 nautical miles.

The major activities planned during the sixth through tenth mission work days include landmark tracking exercises, spacecraft systems exercises, and a multispectral terrain photography experiment for Earth resource studies.

The eleventh work period begins with stowage of onboard equipment and preparations for the SPS

APOLLO 9
Sixth Through Ninth Days

SERVICE PROPULSION BURNS

LANDMARK SIGHTINGS, PHOTOGRAPH
SPECIAL TESTS

deorbit burn 700 miles southeast of Hawaii near the end of the 150th revolution. Splashdown for a 10-day mission will be at 9:46 a.m. EST (238:46:30 GET) in the West Atlantic some 250 miles ESE of Bermuda and 1,290 miles east of Cape Kennedy (30.1 degrees north latitude by 59.9 degrees west longitude).

The Apollo 9 crew and spacecraft will be picked up by the landing platform-helicopter (LPH) USS Guadalcanal. The crew will be airlifted by helicopter the following morning to Norfolk, Va., and thence to the Manned Spacecraft Center, Houston. The spacecraft will be taken off the Guadalcanal at Norfolk, deactivated and flown to the North American Rockwell Space Division plant in Downey, Calif., for postflight analysis.

The Saturn V launch vehicle with the Apollo spacecraft on top stands 363 feet tall. The five first stage engines of Saturn V develop a combined thrust of 7,720,174 pounds at first motion. Thrust increases with

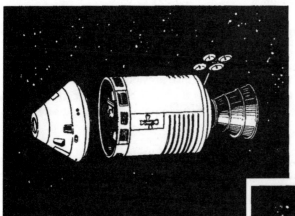

CM / SM SEPARATION

APOLLO 9
Tenth Day

RE-ENTRY

ATLANTIC - SPLASHDOWN

altitude until the total is 9,169,560 pounds an instant before center engine cutoff, scheduled for 2 minutes 14 seconds after liftoff. At that point, the vehicle is expected to be at an altitude of about 26 nm (30 sm, 45 km) and have a velocity of about 5,414 f/sec (1,650 M/sec, 3,205 knots, 3,691 mph). At first stage ignition, the space vehicle will weigh 6,486,915 pounds.

Apollo/Saturn V vehicles were launched Nov. 9, 1967, April 4, 1968, and Dec. 21, 1968, on Apollo missions. The last vehicle carried the Apollo 8 crew, the first two were unmanned.

The Apollo 9 Saturn V launch vehicle is different from the previous Saturn V's in the following aspects:
Dry weight of the first stage has been reduced from 304,000 to 295,600 pounds.
The first stage fueled weight at ignition has been increased from 4,800,000 to 4,946,337 pounds.

Instrumentation measurements in the first stage have been reduced from 891 to 648.

The camera instrumentation electrical power system was not installed on the first stage, and the stage carries neither a film nor television camera system.

The second stage will be somewhat lighter and slightly more powerful than previous S-II's. Maximum vacuum thrust for J-2 engines was increased from 225,000 to 230,000 pounds each. This changed second stage total thrust from 1,125,000 to 1,150,000 pounds. The maximum S-II thrust on this flight is expected to be 1,154,254 pounds.

The approximate dry weight of the S-II has been reduced from 88,000 to 84,600 pounds. The interstage weight was reduced from 11,800 to 11,664 pounds. Weight of the stage fueled has been increased from 1,035,000 to 1,069,114 pounds.

The S-II instrumentation system was changed from research and development to operational, and instrumentation measurements were reduced from 1,226 to 927.

Major differences between the S-IVB used on Apollo 8 and the one for Apollo 9 include:

Dry stage weight decreased from 26,421 to 25,300 pounds. This does not include the 8,081-pound interstage section. Fueled weight of the stage has been decreased from 263,204 to 259,337 pounds.

Stage measurements evolved from research and development to operational status, and instrumentation measurements were reduced from 342 to 280. In the instrument unit, the rate gyro timer, thermal probe, a measuring distributor, tape recorder, two radio frequency assemblies, a source follower, a battery and six measuring racks have been deleted. Instrumentation measurements were reduced from 339 to 221.

During the Apollo 9 mission, communications between the spacecraft and the Mission Control Center, the spacecraft will be referred to as "Apollo 9" and the Mission Control Center as "Houston". This is the procedure followed in past manned Apollo missions.

However, during the periods when the lunar module is manned, either docked or undocked, a modified call system will be used.

Command Module Pilot David Scott in the command module will be identified as "Gumdrop" and Spacecraft Commander James McDivitt and Lunar Module Pilot Russell Schweickart will use the call sign "Spider."

Spider, of course is derived from the bug-like configuration of the lunar module. Gumdrop is derived from the appearance of the command and service modules when they are transported on Earth. During shipment they were wrapped in blue wrappings giving the appearance of a wrapped gumdrop.

(END OF GENERAL RELEASE; BACKGROUND INFORMATION FOLLOWS)

APOLLO 9 COUNTDOWN

The clock for the Apollo 9 countdown will start at T-28 hours, with a six hour built-in-hold planned at T-9 hours, prior to launch vehicle propellant loading.

The countdown is preceded by a pre-count operation that begins some 5½ days before launch. During this period the tasks include mechanical buildup of both the command/ service module and LM, fuel cell activation and servicing and loading of the super critical helium aboard the LM descent stage. A five hour built-in-hold is scheduled between the end of the pre-count and start of the final countdown.

Following are some of the highlights of the final count:

T-28 hrs.	--Official countdown starts
T-27 hrs.	--Install launch vehicle flight batteries (to 23 hrs.)
	--LM stowage and cabin closeout (to 15 hrs.)
T-24 hrs. 30 mins.	--Launch vehicle systems checks (to 18:30 hrs.)
T-20 hrs.	--Top off LM super critical helium (to 17 hrs.)
T-16 hrs.	--Launch vehicle range safety checks (to 15 hrs.)
T-11 hrs. 30 mins.	--Install launch vehicle destruct devices
	--Command/service module pre-ingress operations
T-10 hrs. 30 mins.	--Start mobile service structure move to park site
T-9 hrs.	--Start six hour built-in-hold
T-9 hrs. counting	--Clear blast area for propellant loading
T-8 hrs. 15 mins.	--Launch vehicle propellant loading, three stages (liquid oxygen in first stage; liquid oxygen and liquid hydrogen in second, third stages). Continues through T-3:30 hrs.
T-3 hrs. 10 mins.	--Spacecraft closeout crew on station
T-2 hrs. 40 mins.	--Start flight crew Ingress
T-1 hr. 55 mins.	--Mission Control Center-Houston/ spacecraft command checks
T-1 hr. 50 mins.	--Abort advisory system checks
T-1 hr. 46 mins.	--Space vehicle Emergency Detection System (EDS) test
T-43 mins.	--Retrack Apollo access arm to stand by position (12 degrees)
T-42 mins,	--Arm launch escape system
T-40 mins.	--Final launch vehicle range safety checks
T-30 mins.	--Launch vehicle power transfer test
T-20 mins.	--LM switch over to internal power
T-15 mins.	--Spacecraft to internal power
T-6 mins.	--Space vehicle final status checks
T-5 mins. 30 sec.	--Arm destruct system
T-5 mins.	--Apollo access arm fully retracted
T-3 mins. 10 sec.	--Initiate firing command (automatic sequencer)
T-50 sec.	--Launch vehicle transfer to internal power
T-8.9 sec.	--Ignition sequence start
T-2 sec.	--All engines running
T-0	--Liftoff

*NOTE: Some changes in the above countdown are possible as a result of experience gained in the Countdown Demonstration Test (CDDT) which occurs about 10 days before launch.

SEQUENCE OF EVENTS NOMINAL MISSION

Time from Liftoff

(Hr: Min: Sec)	Date	Time (EST)	Event
00:00:00	Feb. 28	11:00 AM	First Motion
00:00:12			Tilt Initiation
00:01:21			Maximum Dynamic Pressure
00:02:14			S-IC Center Engine Cutoff
00:02:39		11:02:39 AM	S-IC Outboard Engine Cutoff
00:02:40			S-IC/S-II Separation
00:02:42		11:02:42 AM	S-II Ignition
00:03:10			S-II Aft Interstage Separation
00:03:15		11:03:15 AM	Launch Escape Tower Jettison
00:03:21			Initiate IGM
00:08:53		11:08:53 AM	S-II Cutoff
00:08:54			S-II/S-IVB Separation
00:08:57		11:08:57 AM	S-IVB Ignition
00:10:49		11:10:49 AM	S-IVB First Cutoff
00:10:59		11:10:59 AM	Insertion into Earth Parking Orbit
02:34:00		1:34 PM	S-IVB Enters Transposition Attitude
02:43:00		1:43 PM	Spacecraft Separation

02:53:43		1:53 PM	Spacecraft Docking
04:08:57		3:08 PM	Spacecraft Final Separation
04:11:25		3:11 PM	CSM RCS Separation Burn
04:45:41		3:45 PM	S-IVB Reignition
04:46:43		3:46 PM	S-IVB Second Cutoff
04:46:53			S-IVB Insertion into Intermediate Orbit
06:01:40		5:01 PM	SPS Burn No. 1 (Docked)
06:07:04		5:07 PM	S-IVB Reignition
06:11:05		5:11 PM	S-IVB Third Cutoff
06:11:15			S-IVB Insertion into Escape Orbit
06:12:36		5:12 PM	Start S-IVB LOX Dump
06:23:46		5:23 PM	S-IVB LOX Dump Cutoff
06:23:56			Start S-IVB LH2 Dump
06:42:11		5:42 PM	S-IVB LH2 Dump Cutoff
22:12:00	Mar. 1	9:12 AM	SPS Burn No. 2 (Docked)
25:18:30		12:18 PM	SPS Burn No. 3 (Docked)
28:28:00		3:28 PM	SPS Burn No. 4 (Docked)
40:00:00	Mar. 2	3:00 AM	LM Systems Evaluation
46:29:00		9:29 PM	LM TV Transmission
49:43:00		12:43 PM	Docked LM Descent Engine Burn
54:26:16		5:26 PM	SPS Burn No. 5 (Docked)
68:00:00	Mar. 3	7:00 AM	Begin Preps for EVA
73:10:00		12:10 PM	EVA by LM Pilot
75:05:00		2:05 PM	EVA TV
75:20:00		2:20 PM	EVA Ends
89:00:00	Mar. 4	4:00 AM	Rendezvous Preps Begin
93:05:45	Mar. 4	8:05 AM	CSM RCS Separation Burn
93:50:03		8:50 AM	LM Descent Engine Phasing Burn
95:41:48		10:41 AM	LM Descent, Engine Insertion Burn
96:22:00		11:22 AM	LM RCS Concentric Sequence Burn
97:06:28		12:06 PM	LM Ascent Engine Circularization Burn
97:59:21		12:59 PM	LM RCS Terminal Phase Burn
98:31:41		1:31 PM	Terminal Phase Finalization Docking
100:26:00		3:26 PM	LM APS Long Duration Burn
121:59:00	Mar. 5	12:59 PM	SPS Burn No. 6 (CSM only)
169:47:00	Mar. 7	12:47 PM	SPS Burn No. 7
238:10:47	Mar. 10	9:10 AM	SPS Burn No. 8 (Deorbit)
238:41:40		9:41 AM	Main Parachute Deployment
238:46:30		9:46 AM	Splashdown

APOLLO 9 MISSION OBJECTIVES

The Apollo 8 lunar orbit mission in December fully demonstrated that the Apollo command and service modules are capable of operating at lunar distances. But a missing link in flying an Apollo spacecraft to the Moon for a lunar landing is a manned flight in the lunar module.

Apollo 9's primary objective will be to forge that missing link with a thorough checkout in Earth orbit of the lunar module and its systems in a series of tests including maneuvers in which the LM is the active rendezvous vehicle -- paralleling an actual lunar orbit rendezvous.

Apollo 9 will be the most ambitious manned space flight to date, including the Apollo 8 lunar orbit mission. Many more tests are packed into the mission, and most of these deal with the yet-untried lunar module. While the lunar module has been flown unmanned in space (Apollo 5/LM-1, Jan. 22, 1968), the real test of a new spacecraft type comes when it is flown manned. Many of the planned tests will exceed the conditions that will exist in a lunar landing mission.

APOLLO 9 (AS-504) MISSION PROFILE

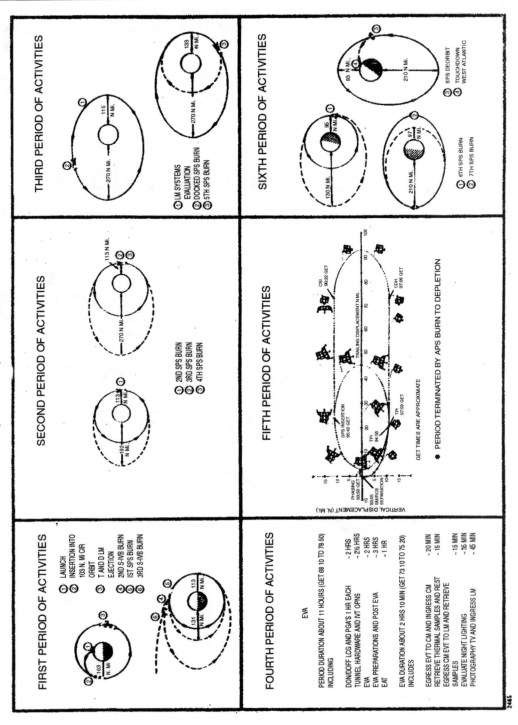

FIRST PERIOD OF ACTIVITIES

① LAUNCH
②② INSERTION INTO 103 N. MI CIR ORBIT
③ T AND D LM EJECTION
④④ 2ND S-IVB BURN 1ST SPS BURN
⑤⑥ 3RD S-IVB BURN

SECOND PERIOD OF ACTIVITIES

① 2ND SPS BURN
② 3RD SPS BURN
③ 4TH SPS BURN

THIRD PERIOD OF ACTIVITIES

① LM SYSTEMS EVALUATION
② DOCKED SPS BURN
③ 5TH SPS BURN

FOURTH PERIOD OF ACTIVITIES

EVA

PERIOD DURATION ABOUT 11 HOURS (GET 68 10 TO 78 50) INCLUDING

DON/DOFF LCG AND PGA'S 1 HR EACH	~ 2 HRS
TUNNEL HARDWARE AND IVT OPNS	~ 2½ HRS
EVA	~ 2 HRS
EVA PREPARATIONS AND POST EVA	~ 3 HRS
EAT	~ 1 HR

EVA DURATION ABOUT 2 HRS 10 MIN (GET 73 10 TO 75 20) INCLUDES

EGRESS EVT TO CM AND INGRESS CM	~ 20 MIN
RETRIEVE THERMAL SAMPLES AND REST	~ 15 MIN
EGRESS CM EVT TO LM AND RETRIEVE SAMPLES	~ 15 MIN
EVALUATE NIGHT LIGHTING	~ 35 MIN
PHOTOGRAPHY TV AND INGRESS LM	~ 45 MIN

FIFTH PERIOD OF ACTIVITIES

GET TIMES ARE APPROXIMATE

● PERIOD TERMINATED BY APS BURN TO DEPLETION

SIXTH PERIOD OF ACTIVITIES

① 6TH SPS BURN
② 7TH SPS BURN
③ SPS DEORBIT
④ TOUCHDOWN WEST ATLANTIC

Although Apollo 9 will be followed by a lunar orbit mission in which the LM descends to 50,000 feet above the Moon's surface, but does not land, there will not be any other long duration burns of the descent engine before the first lunar landing -- possibly on Apollo 11.

Top among mission priorities are rendezvous and docking of the command module and the lunar module. The vehicles will dock twice -- once when the LM is still attached to the S-IVB, and again following the rendezvous maneuver sequence. The dynamics of docking the vehicles can be likened to coupling two freight cars in a railroad switching yard, but using a coupling mechanism built with the precision of a fine watch.

Next in mission priority are special tests of lunar module systems, such as performance of the descent and ascent engines in various guidance control modes and the LM environmental control and electrical power systems that can only get final checkout in space.

The preparations aboard the LM for EVA -- checkout of the LM pilot's extravehicular mobility unit and configuring the LM to support EVA are also of high priority. The Apollo 9 EVA will be the only planned EVA in the Apollo program until the first lunar landing crewmen climb down the LM front leg to walk upon the lunar surface.

CMS/S-IVB ORBITAL OPERATIONS

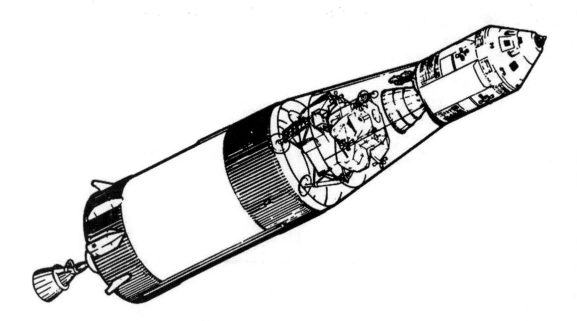

NASA HQ MA69-4177
1-28-69

MISSION TRAJECTORY AND MANEUVER DESCRIPTION

(Note: Information presented herein is based upon a nominal mission and is subject to change prior to the mission or in real time during the mission to meet changing conditions.)

Launch

Apollo 9 is scheduled to be launched at 11 a.m. EST from NASA Kennedy Space Center Launch Complex 39A on a 72-degree azimuth and inserted into a 103 nm (119 sm, 191.3 km) circular Earth orbit by Saturn V launch vehicle No. 504. Insertion will take place at 10 minutes 59 seconds after liftoff.

Transposition and Docking

Following insertion into orbit, the S-IVB third stage maintains an attitude level with the local horizontal while the Apollo 9 crew conducts post-insertion CSM systems checks and prepares for a simulated S-IVB translunar injection restart.

At 2 hours 34 minutes ground elapsed time (GET) the S-IVB enters transposition and docking attitude (15 degree pitch, 35 degree yaw south); and the CSM separates from the S-IVB at 2 hours 43 minutes GET at one fps to about 50 feet separation, where velocity is nulled and the CSM pitches 180 degrees and closes to near the lunar module docking collar for station keeping. Docking is completed at about 2 hours 53 minutes GET and the LM is pressurized with the command module surge tanks and reentry bottles.

LM Ejection and Separation

The lunar module is ejected from the spacecraft/LM adapter by spring devices at the four LM landing gear "knee" attach points. A three-second SM RCS burn separates the CSM/LM for crew observation of the first S-IVB restart.

S-IVB Restarts

After spacecraft separation, the S-IVB resumes a local horizontal attitude for the first J-2 engine restart at 4:45:41 GET. The CSM/LM maintains a separation of about 2,000 feet from the S-IVB for the restart. A second S-IVB restart at 6:07:04 GET followed by propellant dumps place the S-IVB in an Earth escape trajectory and into solar orbit.

TRANSPOSITION, DOCKING AND CSM/LM EJECTION

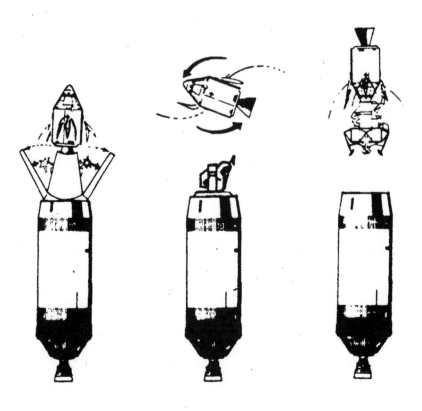

S-IVB SOLO OPERATIONS

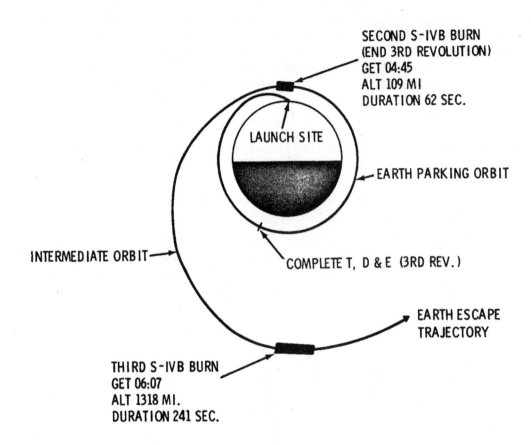

Service Propulsion System (SPS) Burn No. I

A 36.8 fps (11.2 m/sec) docked SPS burn at 6:01:40 GET enhances spacecraft orbital lifetime and demonstrates stability of the CSM digital autopilot. The new orbit is 113 x 131 nm (130 x 151 sm, 290 x 243 km).

Docked SPS Burn No. 2

This burn at 22:12:00 GET reduces CSM weight by 7,355 pounds (3,339 kg) so that reaction control propellant usage in a contingency LM rescue (CSM-active rendezvous) would be lessened; provides continuous SM RCS deorbit capability; tests CSM digital autopilot in 40 percent amplitude stroking range. The burn is mostly out of plane adjusting orbital plane eastward (nominally) and raises apogee to 192 nm (221 sm, 355 km).

Docked SPS Burn No. 3

Another out-of-plane burn at 25:18:30 GET with a velocity change of 2,548.2 fps (776.7 m/sec) further reduces CSM weight and shifts orbit 10 degrees eastward to improve lighting and tracking coverage during rendezvous. The burn will consume 18,637 pounds (8,462 kg) of propellant. Apogee is raised to 270 nm (310 sm, 500 km) and the burn completes the CSM digital autopilot test series. A manual control takeover also is scheduled for this.

DOCKED SPS BURNS

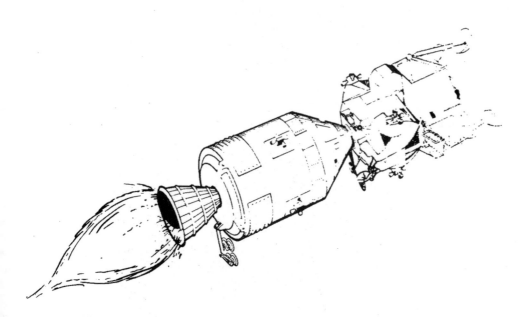

Docked SPS Burn No. 4

A 299.8 fps (91.4 m/sec) out-of-plane burn at 28:28:00 GET shifts the orbit one degree eastward without changing apogee or perigee, but if launch were delayed more than 15 minutes, the burn will be partly in-plane to tune up the orbit for optimum lighting and tracking during rendezvous.

LM Systems Checkout and Power-Up

The commander and lunar module pilot enter the LM through the docking tunnel to power-up the LM and conduct systems checkout, and to prepare for the docked LM descent engine burn. Premission intravehicular transfer timelines will also be evaluated during this period.

DOCKED DPS BURN

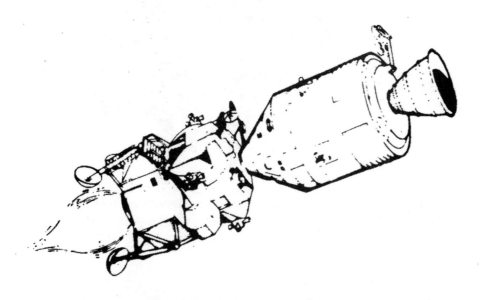

Docked LM Descent Engine Burn

The LM digital autopilot and crew manual throttling of the descent engine will be evaluated in this 1,714.1 fps (522.6 M/sec) descent engine burn at 49:43:00 GET. The burn will shift the orbit eastward 6.7 degrees (for a nominal on-time launch) and will result in a 115 x 270 nm (132 x 310 sm, 213 x 500 km) orbit. After the burn, the LM crew will power down the LM and return to the command module.

Docked SPS Burn No. 5

Apollo 9's orbit is circularized to 133 nm (153 sm, 246 km) by a 550.6 fps (167.8 m/sec) SPS burn at 54:26:16 GET. The 133 nm circular orbit becomes the base orbit for the LM-active rendezvous.

Extravehicular Activity

The commander and the LM pilot will transfer to the LM through the docking tunnel at about 68 hours GET and power up the spacecraft and prepare for EVA. The LM pilot will don the extravehicular mobility unit (EMU), check it out, and leave the LM through the front hatch at 73:10:00 GET. Tethered by a nylon rope, the LM pilot will move with the aid of handrails to the open command module side hatch and place his lower torso into the cabin to demonstrate EVA LM crew rescue. He then will return to the LM "porch" collecting thermal samples from the spacecraft exterior enroute. The LM pilot, restrained by "golden slippers" on the LM porch, photographs various components of the two spacecraft. The commander will pass the LM television camera to the LM pilot who will operate it for about 10 minutes beginning at 75:20:00 GET during a stateside pass. The LM pilot will enter the LM at 75:25:00 GET through the front hatch and the spacecraft will be repressurized. Both crewmen will transfer to the command module after powering down the LM systems.

EXTRAVEHICULAR ACTIVITY

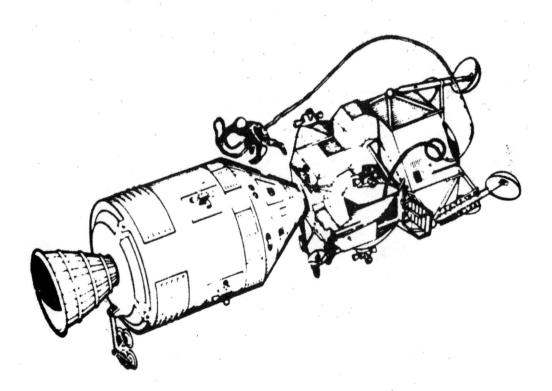

CSM RCS Separation Burn

The commander and the LM pilot transfer to the LM at about 89 hours GET and begin LM power-up and preparations for separation and LM-active rendezvous The first in the series of rendezvous maneuvers is a 5 fps (l.5 m/sec) radially downward CSM RCS burn at 93:05:45 GET which places the CSM in a 131 x 132 nm (151 x 152 sm, 243 x 245 km) equi-period orbit for the "minifootball" rendezvous with a maximum vehicle separation of less than two nautical miles. During the mini-football, the LM radar is checked out and the LM inertial measurement unit (IMU) is aligned.

LM Descent Engine Phasing Burn

An 85 fps (25.9 m/sec) LM descent engine burn radially upward at 93:50:03 GET places the LM in an equi-period 119 x 145 nm (137 x 167 sm, 220 x 268 km) orbit with apogee and perigee about 11.8 miles above and below, respectively, those of the command module. Maximum range following this maneuver is 48 nm and permits an early terminal phase maneuver and rendezvous if the full rendezvous sequence for some reason is no-go. The burn is the only one that is controlled by the 124 abort guidance system with the primary system acting as a backup.

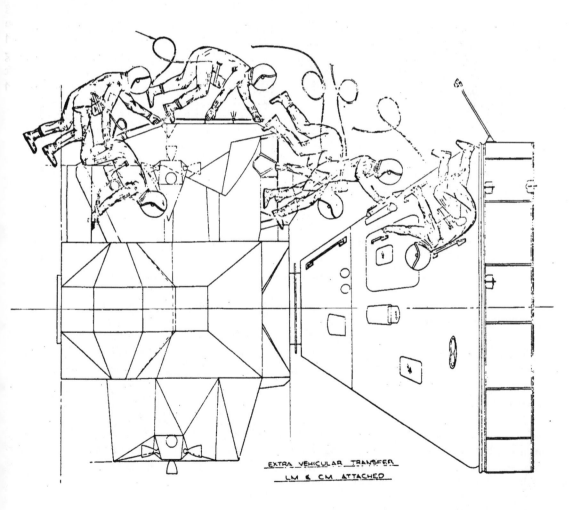

EXTRA VEHICULAR TRANSFER
LM & CM ATTACHED

EVA SCHEDULE (MISSION DAY 4)

*ABOUT 11 HOURS (GET 68:10 TO 78:50) INCLUDING

DON/DOFF LCG AND PGA'S (1 HR. EA.)		2 HRS
TUNNEL HARDWARE AND IVT OPNS		2 ½ HRS
EVA PREPARATIONS AND POST EVA		3 HRS
EAT		1 HR

*EVA DURATION 2 HRS 10 MIN (GET 73:10 TO 75:20)

INCLUDES:

EGRESS, EVT TO CM, AND INGRESS CM	20 MIN.
RETRIEVE THERMAL SAMPLES AND REST	15 MIN.
EGRESS CM, EVT TO LM, AND RETRIEVE SAMPLES	15 MIN.
EVALUATE NIGHT LIGHTING	35 MIN.
PHOTOGRAPHY, TV, AND INGRESS LM	45 MIN.

RENDEZVOUS MANEUVERS AND BACKUP

MANEUVER	SEPARATION	PHASING	INSERTION	CSI	CDH	TPI
GET METHOD DIRECTION DURATION VELOCITY	93:05 CSM/RCS CSM-DOWN 10.9 SEC. 5 FPS	93:50 DPS-AGS UP 25.2 SEC. 85 FPS	95:42 DPS-PGNCS POSIGRADE 24.8 SEC. 40 FPS	96:22 LM/RCS RETROGRADE 30.6 SEC. 38 FPS	97:06 APS-PGNCS RETROGRADE 3.1 SEC. 38 FPS	97:59 LM/RCS POSIGRADE 17.6 SEC. 22 FPS
BACKUP BURNS	NONE REQUIRED		CSM INSERTION - CANCELLING BURN	CSM MIRROR IMAGE BURN	LM-RCS OR CSM MIRROR IMAGE BURN	CSM MIRROR IMAGE BURN
ABORT CAPABILITY	CSM OR LM EXECUTE LINE- OF-SIGHT RANGE AND RANGE RATE CONTROL	CSM OR LM EXECUTE TPI OR LINE-OF- SIGHT RANGE RATE CON- TROL	TWO IMPULSE TECHNIQUE TO REDUCE RETURN TIME TO ABOUT 56 MIN.			DOCKING NOMINALLY AT 99:17

LM Descent Engine Insertion Burn

This burn at 95:41:48 GET of 39.9 fps (9.2 m/sec) is at a 10-percent throttle setting and places the LM in a 142 x 144 nm (163 x 166 sm, 263 x 265 km) near circular orbit and 11 miles above that of the CSM.

CSM Backup Maneuvers

For critical rendezvous maneuvers after the insertion burn, the CM pilot will be prepared to make a "mirror-image" burn of equal velocity but opposite in direction one minute after the scheduled LM maneuver time if for some reason the LM cannot make the maneuver. Such a CSM burn will cause the rendezvous to be completed in the same manner as if the LM had maneuvered nominally.

LM RCS Concentric Sequence Burn (CSI)

A retrograde 37.8 fps (815 m/sec) RCS burn at 96:22:00 GET lowers LM perigee to about 10 miles below that of the CSM. The LM RCS is interconnected to the ascent engine propellant tankage for the burn, and the descent stage is jettisoned prior to the start of the burn.

"D" MISSION RENDEZVOUS PROFILE
(CSM-CENTERED RELATIVE MOTION)

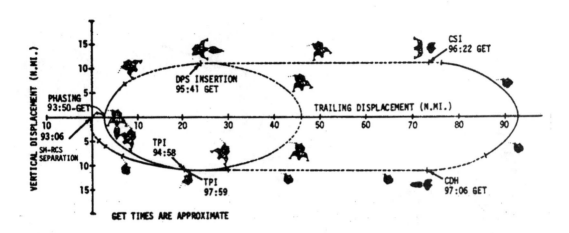

The CSI burn is nominally computed onboard the LM to cause the phase angle at constant delta height (CDH) to result (after CDH is performed) in proper time for terminal phase initiate (TPI).

LM Ascent Engine Circularization Burn (CDH)

The LM ascent stage orbit is circularized by a 37.9 fps (8.5 m/sec) retrograde burn at 97:06:28 GET at LM perigee. At the end of the burn, the LM ascent stage is about 75 miles behind the CSM and in a co-elliptic orbit about 10 miles below that of the CSM. The new LM ascent stage orbit becomes 119 x 121 nm (137 x 139 sm, 220 x 224 km).

LM RCS Terminal Phase Initialization Burn

At 97:59:21 GET, when the LM ascent stage is about 20 nm behind and 10 nm below the CSM, a LM RCS 21.9 fps (6.6 m/sec) burn along the line of sight toward the CSM begins the final rendezvous sequence. Midcourse corrections and braking maneuvers complete the rendezvous and docking is estimated to take place at 98:31:41.

LM Ascent Engine Long-Duration Burn

Following docking and crew transfer back into the CSM, the LM ascent stage is jettisoned and the CSM maneuvers out of plane for separation. The LM ascent engine is ground commanded to burn to depletion at 100:26:00 GET with LM RCS propellant augmented by the ascent tankage crossfeed. The burn will raise LM ascent stage apogee to about 3,200 nm (3,680 am, 5,930 km). The burn is 5,658.5 fps (1,724.7 m/sec).

SPS Burn No. 6

At 121:59:00 GET, the SPS engine is ignited in a 66.1 fps (20.1 m/sec) retrograde burn to lower perigee to 95 nm (109 sm, 176 km) to improve the spacecraft's backup capability to deorbit with the RCS thrusters.

LM ACTIVE RENDEZVOUS

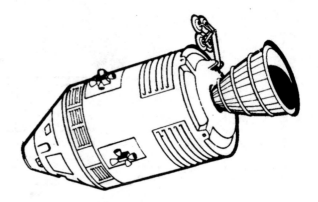

APS BURN TO DEPLETION

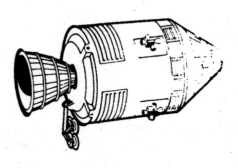

CSM SOLO SPS BURNS

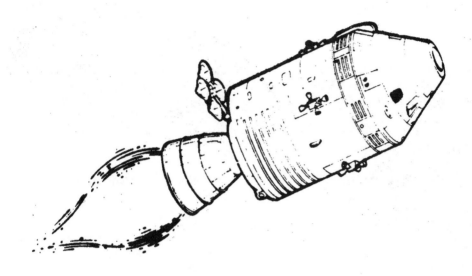

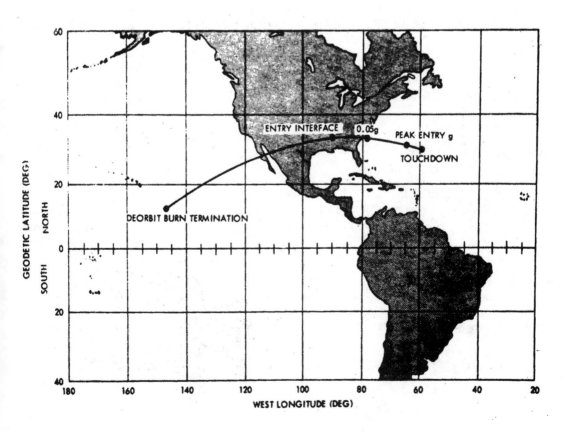

SPS Burn No. 7

Apollo 9 orbital lifetime is extended and RCS deorbit capability improved by this 173.6 fps (52.6 m/sec) posigrade burn at 169:47:00 GET. The burn raises apogee to 210 nm (241 sm, 388 km). The burn also shifts apogee to the southern hemisphere to allow longer free-fall time to entry after a nominal SPS deorbit burn.

SPS Burn No. 8

The SPS deorbit burn is scheduled for 238:10:47 GET and is a 252.9 fps (77 m/sec) retrograde burn beginning about 700 nm southeast of Hawaii.

Entry

After deorbit burn cutoff, the CSM will be yawed 45 degrees out of plane for service module separation. Entry (400,000 feet) will begin about 15 minutes after deorbit burn, and main parachutes will deploy at 238:41:40 GET. Splashdown will be at 238: 46:30 GET at 30.1 degrees north latitude, 59.9 degrees west longitude.

APOLLO 9 MISSION EVENTS

Event	Ground Elapsed Time HRS: MIN: SEC	Velocity Change fps (M/sec)	Purpose (Resultant Orbit)
Insertion	00:10:59	25,245.6(7,389.7)*	Insertion into 103 nm (191.3 km) circular Earth parking orbit
CSM separation, docking	02:43:00	1 (.3)	Hard-mating of CSM and LM docking tunnel
LM ejection	04:08:57	.4 (.12)	Separates LM from S-IVB/SLA
CSM RCS separation burn	04:11:25	3 (.1)	Provides separation prior to S-IVB restart
Docked SPS burn No. 1	06:01:40	36.8 (11.2)	Orbital lifetime, first test of CSM digital autopilot (orbit: 113 x 128 nm)
Docked SPS burn No. 2	22:12:00	849.6 (259)	Reduces CSM weight for LM rescue and RCS deorbit; Second digital autopilot stroking test (113 x 192 nm)
Docked SPS burn No. 3	25:18:30	2,548.2 (776.7)	Reduces CSM weight; completes test of digital autopilot stroking (115 x 270 nm)
Docked SPS burn No. 4	28:28:00	299.8 (91.4)	Out-of-plane burn adjusts orbit for launch delays; no apogee/ perigee change if on-time launch. Late launch: apogee between 130 and 500 nm.
LM Systems evaluation	40:00:00	---	Demonstrate crew intravehicular transfer to LM; power up and checkout LM
Docked LM des. engine burn	49:43:00	1,714.1 (522.6)	Demonstrate LM digital autopilot control, manual throttling of descent engine
Docked SPS burn No. 5	54:26:16	550.6 (167.8)	Circularizes orbit to 133 nm for rendezvous

Event	Ground Elapsed Time HRS: MIN: SEC	Velocity Change fps (M/sec)	Purpose (Resultant Orbit)
Extravehicular activity	73:10:00	---	LM pilot demonstrates EVA transfer from LM to CSM and return; evaluates CM hatch, collects samples on spacecraft exterior, evaluates EVA lighting, performs EVA photography and TV test
CSM RCS separation burn	93:05:45	5.0 (1-5)	Radially downward burn to place CSM in equi-period orbit for mini-football rendezvous (131 x 132)
LM des. engine phasing burn	93:50:03	85 (25.9)	Demonstrates LM abort guidance system, control of descent engine burn, provides radial separation for LM- active rendezvous (LM 119 x 145)
LM des. engine insertion burn	95:41:48	39.9 (9.2)	Adjusts height differential and thus phase angle between LM and CSM at time of CSI maneuver (LM 142 x144)
LM RCS concentric seq. burn	96:22:00	37.8 (8.5)	Retrograde burn adjusts LM perigee to about 10 nm below that of CSM; LM is staged before burn
LM APS circularization burn	97:06:28	37.9 (8-5)	Retrograde burn circularizes LM orbit to about 10 nm below and 75 nm behind and closing on the CSM (LM 119 x 121)
LM RCS terminal phase burn	97:59:21	21.9 (6.6)	LM thrusts along line of sight toward CSM to begin intercept trajectory (LM 121 x 133)
Terminal phase finalization, rendezvous	98:32:41	28.4 (8.6)	Station keeping, braking maneuvers to complete rendezvous (LM/ CSM 131 X 133); docking should occur at 99:18
LM APS long duration burn (unmanned)	100:26:00	5,658.5(1,724.7)	LM ascent engine ground commanded to burn to depletion after CSM undocks; demonstrates ascent engine ability for lunar landing mission profile (LM 131 x 3,200)
SPS burn No. 6 (CSM alone)	121:59:00	66.1 (20.1)	Lowers CSM perigee to enhance RCS deorbit capability (95x 130)
SPS burn No. 7 (CSM alone)	169:47:00	173.6 (52.6)	Extends orbital lifetime and lessens RCS deorbit requirements by raising apogee (97 x 210)
SPS burn No. 8 (CSM alone),	238:10:47	252.9 (77)	Deorbits spacecraft (29 x 210)
Main parachutes deploy	238:41:41	---	
Splashdown	238:46:30		Landing at 30.1 degrees North Latitude x 59.9 degrees West Longitude

* This figure is orbital insertion velocity

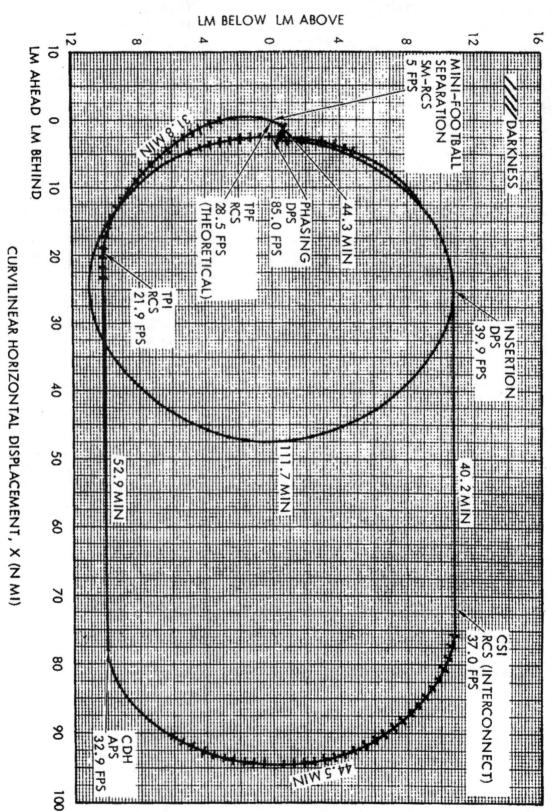

CURVILINEAR VERTICAL DISPLACEMENT, Z (N MI)

LM BELOW LM ABOVE

FLIGHT PLAN SUMMARY

Following is a brief summary of tasks to be accomplished in Apollo 9 on a day-to-day schedule. Apollo 9 work days are not on a 24-hour basis but rather on a variable mission phase and crew activity basis. Rest periods are scheduled at irregular intervals between mission phases.

Launch Day (0-19 hours elapsed time):

CSM systems checkout following insertion into 103 nm orbit

Preparations for transposition, docking and LM ejection

CSM separation, transposition and docking with lunar module

LM pressurization to equal of CSM

LM ejection from spacecraft/LM adapter

Spacecraft evasive maneuver to observe S-IVB restart

Docked posigrade service propulsion system burn raises apogee to 128 nm

Daylight star check and sextant calibration

Second day (19-40 hours GET):

Second SPS docked burn tests digital autopilot stability at 40 per cent of full amplitude gimbal stroke posigrade, raises apogee to 192 nm

Third SPS docked burn tests digital autopilot stability at full stroke; although mostly out-of-plane, burn raises apogee to 270 nm

Fourth docked SPS burn, out-of-plane, orbit remains 115x270 nm

Third day (40-67 GET):

Commander and LM pilot transfer to LM

LM systems checkout

LM alignment optical telescope daylight star visibility check

LM S-Band steerable antenna check

LM platform alignment with combination of known CSM attitudes and voiced data from ground

LM RCS engines hot firing

Docked LM descent engine burn: out of plane, no orbit change

LM crew transfer back into command module

Fifth SPS docked burn circularizes orbit to 133 nm

Fourth day (67-87 GET):

Commander and LM pilot transfer to LM and begin preparations for EVA

CSM and LM depressurized and hatches opened

LM pilot leaves through LM front hatch, mounts camera; command module pilot mounts camera on command module open hatch

LM pilot carries out two-hour EVA in transfer to command module and back to LM for period in "golden slippers" on LM porch; tests LM TV camera

LM pilot enters LM, closes hatch; LM and command module re-pressurized; LM crew returns to command module

Fifth day (87-114 GET):

Commander and LM pilot transfer to LM and prepare for undocking

LM checkout simulating preparations for lunar landing descent

LM rendezvous radar self-test

LM/CSM separation, command module pilot inspects and photographs LM landing gear

LM rendezvous radar look-on with CSM transponder

Descent engine phasing burn places LM in 119 x 145 nm orbit for rendezvous radial separation

LM descent engine posigrade burn inserts LM into 142 x 144 nm orbit for co-elliptic rendezvous maneuver

LM RCS retrograde burn and staging of LM descent stage; LM ascent stage now in 120 x 139 nm orbit

LM ascent engine co-elliptic burn places LM in 119 x 121 nm orbit; LM orbit now constant about 10 nm below CSM

LM RCS terminal phase initiation burn along line of sight toward CSM, followed by midcourse correction burns

Rendezvous and docking

LM crew transfer to command module

LM jettisoned and ascent engine ground-commanded for burn to depletion

Sixth day (114-139 GET):

Sixth SPS burn lowers CSM perigee to 95 nm

Seventh day (139-162 GET):

Landmark tracking over U.S. and South Atlantic

Eighth day (162-185 GET):

Seventh SPS burn raises apogee to 210 nm

Ninth day (185-208 GET):

No major mission activities planned

Tenth day (208-231 GET):

No major mission activities planned

Eleventh day (231 GET to splashdown):

Entry preparations

SPS fps retrograde deorbit burn

CM/SM separation, entry and splashdown

APOLLO 9 ALTERNATE MISSIONS

Any of the several alternate mission plans possible for Apollo 9 will focus upon meeting the most lunar module test objectives. Depending upon when in the mission timeline a failure occurs, and what the nature of the failure is, an alternate mission plan will be chosen in real time.

Failures which would require a shift to an alternate mission plan are cross-matched against mission timeline periods in an alternate mission matrix from which the flight control team in Mission Control Center would choose an alternate promising the maximum in mission objectives met.

The functional failure side of the alternate mission matrix has 14 possible failures. They are: Early S-IVB cutoff, forcing an SPS contingency orbit insertion (COI) and a CSM only alternate mission; LM cannot be ejected from the spacecraft/ LM adapter (SLA) ; SPS will not fire; problem with CSM lifetime; failure of either CSM coolant loop; unsafe descent stage; LM descent propulsion engine inoperable; extravehicular transfer takes longer than 15 minutes; loss of LM primary guidance, navigation and control system (PGNCS); loss of LM primary coolant loop; electrical power problem in LM descent stage; electrical problem in LM ascent stage; LM rendezvous radar failure; and loss of LM abort guidance system.

Seven basic alternate missions have been outlined, each of which has several possible sub-alternates stemming from when a failure occurs and how many of the mission's objectives have been accomplished.

The basic alternate missions are summarized as follows:

Alternate A: (No lunar module because of SPS contingency orbit insertion or failure of LM to eject from SLA.) Full-duration CSM only mission with the scheduled eight SPS burns on the nominal timeline.

Alternate B: (No SPS, lifetime problems on CSM and LM.) Would include transposition, docking and extraction of LM; LM systems evaluation, docked DPS burn, EVA, station keeping with ascent stage, long ascent engine burn, and RCS deorbit.

Alternate C: (Unsafe descent stage, EVA transfer runs overtime.) Crew would perform EVA after separating descent stage, the long ascent engine burn would be performed and the balance of the mission continue along the nominal timeline.

Alternate D: (CSM and/or LM lifetime problems, failure of either CSM coolant loop.) Mission plan would be reshaped to include transposition.. docking and LM extraction, LM systems evaluation, docked descent engine burn, staging and long ascent engine burn, and continuation along nominal timeline.

Alternate E: (unsafe descent stage, descent engine failure, LM primary coolant loop failure electrical problems in either LM stage, PGNCS failure, rendezvous radar failure, or loss of LM abort guidance system.) Four possible modified rendezvous plans are in this alternate, each depending upon the nature of the systems failure and when it takes place. The modified rendezvous are: station keeping, mini-football rendezvous, football rendezvous and CSM active rendezvous.

Alternate F: (Failure of LM primary guidance, navigation and control system.) This alternate would delete docked descent engine burn, long ascent engine burn, but would include SPS burn No. 5, EVA station keeping and docking with LM ascent stage, and a CSM-active rendezvous.

Alternate G: (Failure of LM primary coolant loop or descent engine inoperable.) The docked descent engine burn and the LM-active rendezvous is deleted in this alternate, but the long ascent engine burn is done after undocking and the mission then follows the nominal timeline.

D Mission	ALTERNATE MISSIONS
ALTERNATE	DESCRIPTION
A.	Contingency: LM fails or cannot be ejected from SLA. Perform CSM-only mission.
B.	Contingency: Curtailed CSM systems performance. Accomplish priority objectives on accelerated time scale.
C.	Contingency: DPS cannot be ignited, or fails during docked burn. Perform EVA and long APS burn. Accomplish CSM objectives.
D.	Contingency: Loss of CSM ECS coolant loop. Depending on time of occurrence, conduct LM evaluation, execute docked DPS burn, station keeping, and long APS burn.
E.	Contingency: Various LM subsystem failures. Perform rendezvous as modified in real-time.
F.	Contingency: Loss of PGNCS. Delete docked DPS burn and long APS burn, perform EVA, and substitute CSM active rendezvous.
G.	Contingency: LM primary coolant loop failure. Delete docked DPS burn. Perform EVA and long APS burn.

ABORT MODES

The Apollo 9 mission can be aborted at any time during the launch phase or during later phases after a successful insertion into Earth orbit.

Abort modes can be summarized as follows:

Launch phase -

Mode I - Launch escape tower propels command module away from launch vehicle. This mode is in effect from about T-20 minutes when LES is armed until LES jettison at 3:16 GET and command module landing point can range from the Launch Complex 39A area to 520 nm (600 sm, 964 km) downrange.

Mode II - Begins when LES is jettisoned and runs until the SPS can be used to insert the CSM into a safe

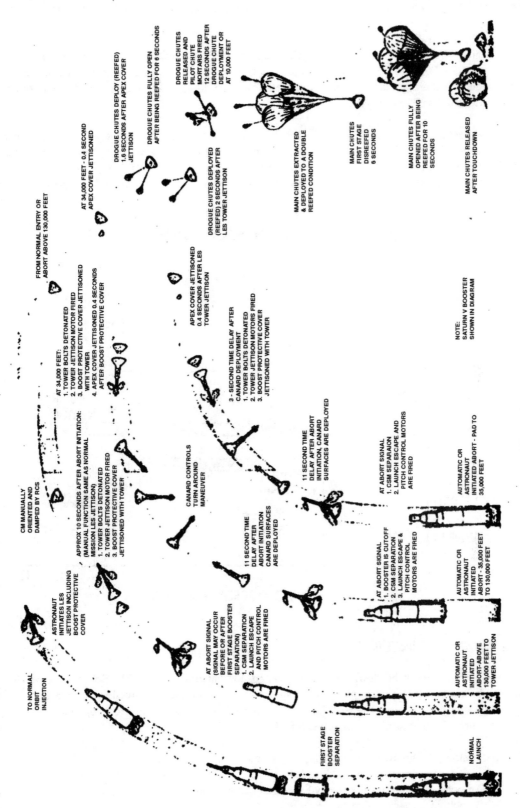

LES ABORT DIAGRAM

orbit (9:22 GET) or until landing points threaten the African coast. Mode II requires manual separation, entry orientation and full-lift entry with landing between 400 and 3,200 nm (461- 3,560 sm, 741-5,931 km) downrange.

Mode III - Begins when full-lift landing point reaches 3,200 nm (3,560 sm, 5,931 km) and extends through orbital insertion. The CSM would separate from the launch vehicle, and if necessary, an SPS retrograde burn would be made, and the command module would be flown half-lift to entry and landing at approximately 3,350 nm (3,852 sm, 6,197 km) downrange.

Mode IV and Apogee Kick - Begins after the point the SPS could be used to insert the CSM into an Earth parking orbit from about 9:22 GET. The SPS burn into orbit would be made two minutes after separation from the S-IVB and the mission would continue as an Earth orbit alternate. Mode IV is preferred over Mode III. A variation of Mode IV is the Apogee Kick in which the SPS would be ignited at first apogee to raise perigee for a safe orbit.

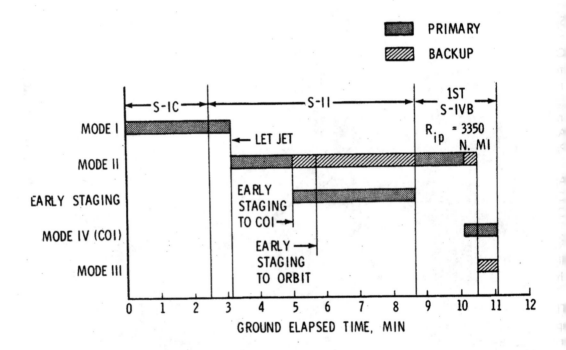

MISSION D LAUNCH ABORT TIMELINE

APOLLO 9 GO/NO-GO DECISION POINTS

Like Apollo 8, Apollo 9 will be flown on a step-by-step commit point or Go/No-Go basis in which the decision will be made prior to each major maneuver whether to continue the mission or to switch to one of the possible alternate missions. The Go/No-Go decisions will be based upon the joint opinions of the flight crew and the flight control teams in Mission Control Center.

Go/No-Go decisions will be made prior to the following events:

1.   Launch phase Go/No-Go at 9 min. 40 sec. GET for orbit insertion.
2.   S-IVB orbit coast period after S-IVB cutoff.
3.   Continue mission past preferred target point 2-1 to target point 6-4.
4.   Transposition, docking and LM extraction.
5.   S-IVB orbital maneuvers.

6.   Service propulsion system maneuvers.
7.   Continuing the mission past target point 6-4.
8.   Daily for going past the West Atlantic target point.
9.   Crew intravehicular transfer to LM.
10.   Docked descent engine burn.
11.   Extravehicular activity.
12.   CSM/LM undocking.
13.   Separation maneuver.
14.   Phasing maneuver.
15.   Insertion maneuver.
16.   LM staging.
17.   Final LM separation and unmanned APS burn.

RECOVERY OPERATIONS

The primary landing point for Apollo 9 is in the West Atlantic at 59.9 degrees West Longitude by 30.1 degrees North Latitude for a nominal full-duration mission. Prime recovery vessel is the helicopter landing platform USS Guadalcanal.

Splashdown for a nominal mission launched on time at 11 a.m. EST, Feb. 28 will be at 9:46 a.m. EST, Mar. 10.

Other ships along the launch-phase ground track, in addition to the Guadalcanal, will be the Apollo instrumentation ship Vanguard and the destroyer USS Chilton off the west coast of Africa. Ships on station in Pacific contingency landing areas include one vessel in the West Pacific and two in the mid-Pacific.

In addition to surface vessels deployed in the launch abort area and the primary recovery vessel in the Atlantic, 16 HC-130 aircraft will be on standby at eight staging bases around the Earth: Tachikawa, Japan; Pago Pago, Samoa; Hawaii; Bermuda; Lajes, Azores; Ascension Island; Mauritius and Panama Canal Zone.

Apollo 9 recovery operations will be directed from the Recovery Operations Control Room in the Mission Control Center and will be supported by the Atlantic Recovery Control Center, Norfolk, Va.; Pacific Recovery Control Center, Kunia, Hawaii; and control centers at Ramstein, Germany and Albrook AFB, Canal Zone.

Following splashdown and crew and spacecraft recovery, the Guadalcanal will steam toward Norfolk, Va. The flight crew will be flown by helicopter to Norfolk the morning after recovery from whence they will fly to the Manned Spacecraft Center, Houston.

The spacecraft will be taken off at Norfolk upon the Guadalcanal's arrival and undergo deactivation for approximately five days. It then will be flown aboard a C-133B aircraft to Long Beach, Calif., and thence trucked to the North American Rockwell Space Division plant in Downey, Calif., for post flight analysis.

PHOTOGRAPHIC EQUIPMENT

Apollo 9 will carry two 70mm standard and one super wide-angle Hasselblad still cameras and two 16 mm Maurer sequence cameras. Film magazines for specific mission photographic objectives are carried for each camera.

The Standard Hasselblad cameras are fitted with 80 mm f/2.8 to f/22 Zeiss Planar lenses, and the Super wide-angle Hasselblad is fitted with a 38mm f/4-5 to f/22 Zeiss Biogon lens. The Maurer sequence cameras have bayonet-mount 75mm f/2-5, 18mm f/2 and 5mm f/2 interchangeable lenses available to the crew.

Hasselblad shutter speeds are variable from 1 sec. to 1/500 sec., and sequence camera frame rates of 1,6,12 and 24 frames per-second can be selected.

Film emulsions have been chosen for each specific photographic task. For example, a medium speed color reversal film will be used for recording docking, EVA and rendezvous and a high-speed color film will be used for command module and lunar module cabin interior photography.

Camera accessories carried aboard Apollo 9 include mounting brackets, right-angle mirror attachments, haze filter, an exposure-measuring spotmeter, a ringsight common to both types of camera, an EVA camera tether and a sequence camera remote controller for EVA photography. Power cables also are included.

APOLLO 9 EXPERIMENT

SO65 Experiment - Multispectral Photography

This experiment, being flown for the first time, is designed to obtain multispectral photography from space over selected land and ocean areas.

Equipment for the experiment consists of four Model 500-EL Hasselblad cameras operated by electric motors, installed in a common mount and synchronized for simultaneous exposure. The mount is installed in the command module hatch window during photographic operations and the spacecraft will be oriented to provide vertical photography. A manual introvolometer is used to obtain systematic overlapping (stereo) photography.

Each camera has a standard 80 millimeter focal length lens and a single film magazine containing from 160 to 200 frames.

Film-filter combinations for the cameras are similar to the ones presently planned for the Earth Resources Technology Satellite (ERTS-A) payload.

Present plans for Earth photography emphasize coverage of the Southwest U.S., where ground information is more readily available. Other areas of high interest include Mexico and Brazil. The domestic area of interest includes Tucson, El Paso, Dallas/Ft. Worth and the Welaco Agricultural Experiment Station in Southwest Texas.

A photographic operations room will be maintained around the clock by the Manned Spacecraft Center's Earth Resources Division. Direct access to the mission controller will provide in-flight reprogramming of photographic coverage to help optimize photographic coverage by considering weather and operational conditions as the mission progresses.

A meeting will be held at the Manned Spacecraft Center as soon as possible after the mission to allow approximately 20 participating scientific investigators and participating user agency representatives to review the photography and get briefed on the details of the experiment. It is estimated that it will take approximately two weeks for the photographic technology laboratory to make the high quality film duplicates that will be provided to each participant at this meeting. The participants will be asked to provide a report within 90 days on their preliminary experimentation and these reports will be compiled and published by NASA.

Films and filters and spectral ranges for the experiment are:

1) Infrared Aerographic film with an 89B filter, 700 mu to 900 mu, to provide narrow-band infrared data for comparison with the responses obtained in the visible region of the electromagnetic spectrum by the other cameras in the experiment.

2) Color IR with Wratten 15 filter, 510 mu to 900 mu region, to take advantage of plant reflectance in the near IR and to provide for maximum differentiation between natural and cultural features.

3) Panatomic-X with 25A filter, 580 mu into the IR region, to provide imagery of value in differentiating various types of land use and in enhancing high contrast objects such as clouds.

) Panatomic-X with 58 filter, 480 mu to 620 mu region, to provide for maximum penetration of lakes and coastal water bottom topography.

The principal investigator is Dr. Paul D. Lowman, Jr., of the NASA Goddard Space Flight Center, Greenbelt.

APOLLO 9 ONBOARD TELEVISION

A lunar television camera of the type that will transmit a video signal back to Earth during Apollo lunar landing missions will be stowed aboard LM-3

Two television transmissions are planned for Apollo 9 - one during the first manning and systems checkout of the LM, and the other during Schweickart's EVA. The first TV pass will be a test with the camera simply warmed up and passively transmitting during the systems checkout, and will last some seven minutes (46:27 - 46:34 GET) during a pass over the MILA tracking station.

After Schweickart has transferred EVA from the LM to the command module and back and is restrained by the "golden slippers" on the LM porch, McDivitt will pass the TV camera out to him for a 10-minute pass over the Goldstone and MILA stations (75:05 - 75:15 GET).

The video signal is transmitted to ground stations by the LM S-Band transmitter. Goldstone, Calif. and Merritt Island, Fla., are the two MSFN stations equipped for scan conversion and output to the Mission Control Center, although other MSFN stations are capable of recording the TV signal at the slow scan rate.

The lunar television camera weighs 7.25 pounds and draws 6.5 watts of 24-32 volts DC power. Scan rate is 10 frames/sec. at 320 lines/frame. The camera body is 10.6 inches long, 6.5 inches wide and 3.4 inches deep. The bayonet lens mount permits lens changes by a crewman in a pressurized suit. Lenses for the camera include a lunar day lens, lunar night lens, a wide-angle lens and a 100mm telephoto lens. The wide-angle, and lunar day lens will be carried with the Apollo 9 camera.

A tubular fitting on the end of the electrical power cable which plugs into the bottom of the camera serves as a handgrip.

The Apollo lunar television camera is built by Westinghouse Electric Corporation Aerospace Division, Baltimore, Md. TV cameras carried on Apollo 7 and 8 were made by RCA.

COMMAND AND SERVICE MODULE STRUCTURE SYSTEMS

The Apollo spacecraft for the Apollo 9 mission is comprised of a Command Module 104, Service Module 104, Lunar Module 3, a spacecraft-lunar module adapter (SLA) 12 and a launch escape system. The SLA serves as a mating structure between the instrument unit atop the S-IVB stage of the Saturn V launch vehicle and as a housing for the lunar module.

Launch Escape System (LES) -- Propels command module to safety in an aborted launch. It is made up of an open-frame tower structure mounted to the command module by four frangible bolts, and three solid propellant rocket motors: a 147,000 pound-thrust launch escape system motor, a 2,400 pound-thrust pitch control motor and a 31,500 pound-thrust tower jettison motor. Two canard vanes near the top deploy to turn the command module aerodynamically to an attitude with the heat-shield forward. Attached to the base of the launch escape tower is a boost protective cover composed of glass, cloth and honeycomb, that protects the command module from rocket exhaust gases from the main and the jettison motor. The system is 33 feet tall, four feet in diameter at the base and weighs 8,848 pounds.

Command Module (CM) Structure -- The basic structure of the command module is a pressure vessel encased in heat-shields, cone-shaped 12 feet high, base diameter of 12 feet 10 inches, and launch weight 12,405 pounds. The command module consists of the forward compartment which contains two negative pitch reaction control engines and components of the Earth landing system; the crew compartment, or inner

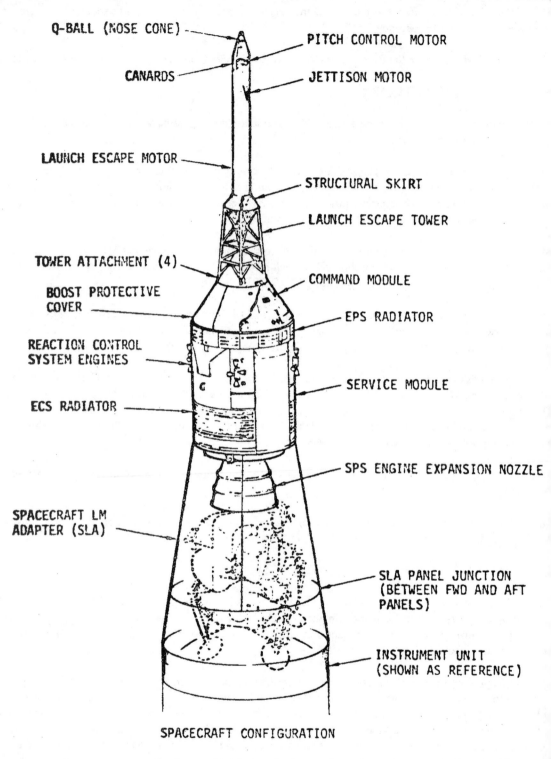

Q-BALL (NOSE CONE)

PITCH CONTROL MOTOR

CANARDS

JETTISON MOTOR

LAUNCH ESCAPE MOTOR

STRUCTURAL SKIRT

LAUNCH ESCAPE TOWER

TOWER ATTACHMENT (4)

COMMAND MODULE

BOOST PROTECTIVE COVER

EPS RADIATOR

REACTION CONTROL SYSTEM ENGINES

ECS RADIATOR

SERVICE MODULE

SPS ENGINE EXPANSION NOZZLE

SPACECRAFT LM ADAPTER (SLA)

SLA PANEL JUNCTION (BETWEEN FWD AND AFT PANELS)

INSTRUMENT UNIT (SHOWN AS REFERENCE)

SPACECRAFT CONFIGURATION

pressure vessel, containing crew accommodations, controls and displays, and spacecraft systems; and the aft compartment housing ten reaction control engines and propellant tankage. Heat-shields around the three compartments are made of brazed stainless steel honeycomb with an outer layer of phenolic epoxy resin a an ablative material. Heat-shield thickness, varying according to heat loads, ranges from 0.7 inches (at th apex) to 2.7 inches on the aft side. The spacecraft inner structure is of aluminum alloy sheet aluminum honeycomb bonded sandwich ranging in thickness from 0.25 inches thick at forward access tunnel to 1. inches thick at base. CSM 104 and LM-3 will carry for the first time the probe and drogue docking hardwa The probe assembly is a folding coupling and impact attenuating device mounted on the CM tunnel tha

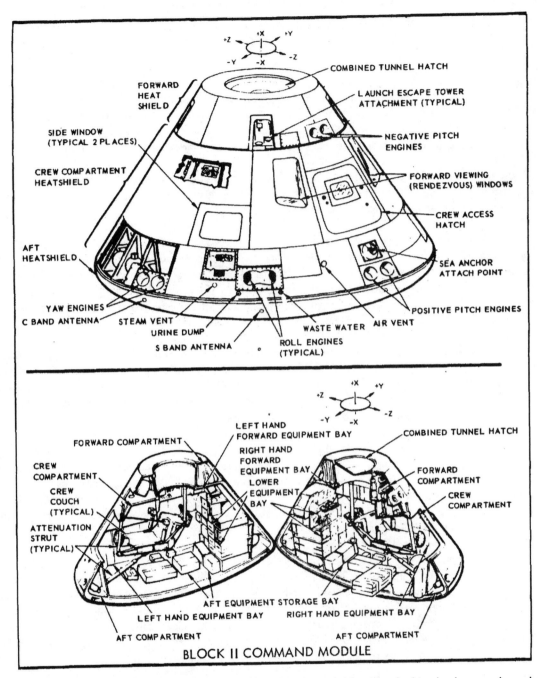

BLOCK II COMMAND MODULE

mates with a conical drogue mounted on the LM docking tunnel. After the docking latches are dogged down following a docking maneuver, both the probe and drogue assemblies are removed from the vehicle tunnels and stowed to allow free crew transfer between the CSM and LM.

Service Module (SM) Structure -- The service module is a cylinder 12 feet 10 inches in diameter by 22 feet long. For the Apollo 9 mission, it will weigh 36,159 Pounds (16,416.2 kg) at launch. Aluminum honeycomb panes one inch thick form the outer skin, and milled aluminum radial beams separate the interior into six sections containing service propulsion system and reaction control fuel-oxidizer tankage, fuel cells, cryogenic oxygen and hydrogen, and onboard consumables.

Spacecraft LM adapter (SLA) Structure -- The spacecraft LM adapter is a truncated cone 28 feet long tapering from 260 inches diameter at the base to 154 inches at the forward end at the service module mating line.

Aluminum honeycomb 1.75 inches thick is the stressed-skin structure for the spacecraft adapter. The SLA weighs 4,107 pounds.

CSM Systems

Guidance, Navigation and Control System (GNCS) Measures and controls spacecraft position attitude and velocity, calculates trajectory, controls spacecraft propulsion system thrust vector and displays abort data. The Guidance System consists of three subsystems: Inertial, made up of an inertial measuring unit and associated power and data components; Computer which processes information to or from other components; and Optics, including scanning telescope, sextant for celestial and/or landmark spacecraft navigation.

Stabilization and Control System (SCS) -- Controls spacecraft rotation, translation and thrust vector and provides displays for crew-initiated maneuvers; backs up the guidance system. It has three subsystems; attitude reference, attitude control and thrust vector control.

Service Propulsion System (SPS) -- Provides thrust for large spacecraft velocity changes through a gimbal-mounted 20,500 pound-thrust hypergolic engine using nitrogen tetroxide oxidizer and a 50-50 mixture of unsymmetrical dimethyl hydrazine and hydrazine fuel. Tankage of this system is in the service module. The system responds to automatic firing commands from the guidance and navigation system or to manual commands from the crew. The engine provides a constant thrust rate. The stabilization and control system gimbals the engine to fire through the spacecraft center of gravity.

Reaction Control System (RCS) --The Command Module and the Service Module each has its own independent system, the CM RCS and the SM RCS respectively. The SM RCS has four identical RCS "quads" mounted around the SM 90 degrees apart. Each quad has four 100 pound-thrust engines, two fuel and two oxidizer tanks and a helium pressurization sphere. The SM RCS provides redundant spacecraft attitude control through cross-coupling logic inputs from the Stabilization and Guidance Systems.

Small velocity change maneuvers can also be made with the SM RCS. The CM RCS consists of two independent six-engine subsystems of six 94 pounds-thrust engines each. Both subsystems are activated after separation from the SM: one is used for spacecraft attitude control during entry. The other serves in standby as a backup. Propellants for both CM and SM RCS are monomethyl hydrazine fuel and nitrogen tetroxide oxidizer with helium pressurization. These propellants are hypergolic, i.e., they burn spontaneously when combined without need for an igniter.

Electrical Power System (EPS) -- Consists of three, 31-cell Bacon - type hydrogen-oxygen fuel cell power plants in the service module which supply 28-volt DC power, three 28-volt DC zinc silver oxide main storage batteries in the command module lower equipment bay, and three 115-200 volt 400 hertz three-phase AC inverters powered by the main 28-volt DC bus. The inverters are also located in the lower equipment bay. Cryogenic hydrogen and oxygen react in the fuel cell stacks to provide electrical power, potable water and heat. The command module main batteries can be switched to fire pyrotechnics in an emergency. A battery charger restores selected batteries to full strength as required with power from the fuel cells.

Environmental Control System (ECS) -- Controls spacecraft atmosphere, pressure and- temperature and manages water. In addition to regulating cabin and suit gas pressure, temperature and humidity, the system removes carbon dioxide, odors and particles, and ventilates the cabin after landing. It collects and stores fuel cell potable water for crew use, supplies water to the glycol evaporators for cooling, and dumps surplus water overboard through the urine dump valve. Proper operating temperature of electronics and electrical equipment is maintained by this system through the use of the cabin heat exchangers, the space radiators and the glycol evaporators.

Telecommunications System -- Provides voice, television telemetry, and data and tracking and ranging between the spacecraft and earth, between the command module and the lunar module and between the spacecraft and the extravehicular astronaut. It also provides intercommunications between astronauts. The telecommunications system consists of pulse code modulated telemetry for relaying to Manned Space Flight

SERVICE MODULE
BLOCK II

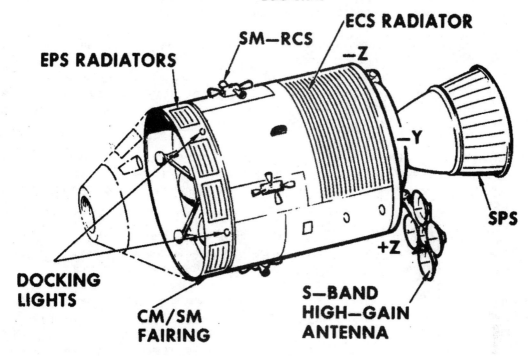

EPS RADIATORS

SM—RCS

ECS RADIATOR

—Z

—Y

SPS

DOCKING LIGHTS

CM/SM FAIRING

S—BAND HIGH—GAIN ANTENNA

+Z

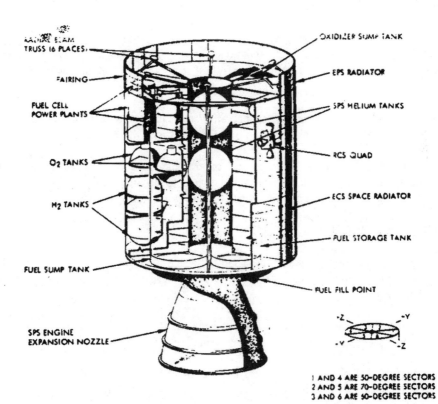

MAIN BEAM TRUSS (6 PLACES)

FAIRING

FUEL CELL POWER PLANTS

O₂ TANKS

H₂ TANKS

FUEL SUMP TANK

SPS ENGINE EXPANSION NOZZLE

OXIDIZER SUMP TANK

EPS RADIATOR

SPS HELIUM TANKS

RCS QUAD

ECS SPACE RADIATOR

FUEL STORAGE TANK

FUEL FILL POINT

1 AND 4 ARE 50-DEGREE SECTORS
2 AND 5 ARE 70-DEGREE SECTORS
3 AND 6 ARE 50-DEGREE SECTORS

Network stations data on spacecraft systems and crew condition, VHF/AM voice, and unified S-Band tracking transponder, air-to-ground voice communications, onboard television (not installed on CM 104) and a VHF recovery beacon. Network stations can transmit to the spacecraft such items as updates to the Apollo guidance computer and central timing equipment, and real-time commands for certain onboard functions. More than 300 CSM measurements will be telemetered to the MSFN.

The high-gain steerable S-Band antenna consists of four, 31-inch-diameter parabolic dishes mounted on a folding boom at the aft end of the service module. Nested alongside the service propulsion system engine nozzle until deployment, the antenna swings out at right angles to the spacecraft longitudinal axis, with the boom pointing 52 degrees below the heads-up horizontal. Signals from the ground stations can be tracked either automatically or manually with the antenna's gimballing system. Normal S-Band voice and uplink/downlink communications will be handled by the omni and high-gain antennas.

Sequential System -- Interfaces with other spacecraft systems and sub systems to initiate time critical functions during launch, docking maneuvers, pre-orbital aborts and entry portions of a mission. The system also controls routine spacecraft sequencing such as service module separation and deployment of the Earth landing system.

Emergency Detection System (EDS) -- Detects and displays to the crew launch vehicle emergency conditions, such as excessive pitch or roll rates or two engines out, and automatically or manually shuts down the booster and activates the launch escape system functions until the spacecraft is in orbit.

Earth Landing System (ELS) -- Includes the drogue and main parachute system as well as post-landing recovery aids. In a normal entry descent, the command module forward heat shield is jettisoned at 24,000 feet, permitting mortar deployment of two reefed 16.5 foot diameter drogue parachutes for orienting and decelerating the spacecraft. After disreef and drogue release, three pilot mortar deployed chutes pull out the three main 83.3 foot diameter parachutes with two-stage reefing to provide gradual inflation in three steps. Two main parachutes out of three can provide a safe landing.

Recovery aids include the uprighting system, swimmer interphone connections, sea dye marker, flashing beacon, VHF recovery beacon and VHF transceiver. The uprighting system consists of three compressor-inflated bags to upright the spacecraft if it should land in the water apex down (Stable II position).

Caution and Warning System -- Monitors spacecraft systems for out-of-tolerance conditions and alerts crew by visual and audible alarms so that crewmen may trouble-shoot the problem.

Controls and Displays -- Provide readouts and control functions of all other spacecraft systems in the command and service modules. All controls are designed to be operated by crewmen in pressurized suits. Displays are grouped by system according to the frequency the crew refers to them.

LUNAR MODULE STRUCTURES, SYSTEMS

The lunar module is a two-stage vehicle designed for space operations near and on the Moon or in Earth orbit developmental missions such as Apollo 9. The LM is incapable of reentering the atmosphere and is, in effect, a true spacecraft.

Joined by four explosive bolts and umbilicals, the ascent and descent stages of the LM operate as a unit until staging, when the ascent stage functions as a single spacecraft for rendezvous and docking with the CSM.

Three main sections make up the ascent stage: the crew compartment, midsection and aft equipment bay. Only the crew compartment and midsection can be pressurized (4.8 Psig; 337.4 gm/sq cm) as part of the LM cabin; all other sections of the LM are unpressurized. The cabin volume is 235 cubic feet (6-7 cubic meters).

Structurally, the ascent stage has six substructural areas: crew compartment, midsection, aft equipment bay, thrust chamber assembly cluster supports, antenna supports and thermal and micrometeoroid shield.

The cylindrical crew compartment is a semimonocoque structure of machined longerons and fusion-welded aluminum sheet and is 92 inches (2.35 m) in diameter and 42 inches (1.07 m) deep. Two flight stations are equipped with control and display panels, armrests, body restraints, landing aids, two front windows, an overhead docking window and an alignment optical telescope in the center between the two flight stations.

Two triangular front windows and the 32-inch (.81 m) square inward-opening forward hatch are in the crew compartment front face.

External structural beams support the crew compartment and serve to support the lower interstage mounts at their lower ends. Ring-stiffened semimonocoque construction is employed in the midsection, with chem-milled aluminum skin over fusion welded longerons and stiffeners. Fore-and-aft beams across the top of the midsection join with those running across the top of the cabin to take all ascent stage stress loads and, in effect, isolate the cabin from stresses.

The ascent stage engine compartment is formed by two beams running across the lower midsection deck and mated to the fore and aft bulkheads. Systems located in the midsection include the LM guidance computer, the power and servo assembly, ascent engine propellant tanks, RCS propellant tanks, the environmental control system, and the waste management section.

A tunnel ring atop the ascent stage meshes with the command module latch assemblies. During docking the ring and clamps are aligned by the LM drogue and the CSM probe.

The docking tunnel extends downward into the midsection 16 inches (40 cm). The tunnel is 32 inches (.81 cm) in diameter and is used for crew transfer between the CSM and LM by crewmen in either pressurized or unpressurized extravehicular mobility units (EMU). The upper hatch on the inboard end of the docking tunnel hinges downward and cannot be opened with the LM pressurized.

A thermal and micrometeoroid shield of multiple layers of mylar and a single thickness of thin aluminum skin encases the entire ascent stage structure.

The descent stage consists of a cruciform load-carrying structure of two pairs of parallel beams, upper and lower decks, and enclosure bulkheads -- all of conventional skin-and-stringer aluminum alloy construction. The center compartment houses the descent engine, and descent propellant tanks are housed in the four square bays around the engine.

Four-legged truss outriggers mounted on the ends of each pair of beams serve as SLA attach points and as "knees" for the landing gear main struts.

Triangular bays between the main beams are enclosed into quadrants housing such components as the ECS water tank, helium tanks, descent engine control assembly of the guidance, navigation and control subsystem, ECS gaseous oxygen tank and batteries for the electrical power system. Like the ascent stage, the descent stage is encased in a mylar and aluminum alloy thermal and micrometeoroid shield.

The LM external platform, or "porch," is mounted on the forward outrigger just below the forward hatch. A ladder extends down the forward landing gear strut from the porch for crew lunar surface operations. Foot restraints ("golden slippers") have been attached to the LM-3 porch to assist the lunar module pilot during EVA photography. The restraints face the LM hatch.

In a retracted position until after the crew mans the LM, the landing gear struts are explosively extended to provide lunar surface landing impact attenuation. The main struts are filled with crushable aluminum honeycomb for absorbing compression loads, Footpads 37 inches (.95 m) in diameter at the end of each landing gear provide vehicle "flotation" on the lunar surface.

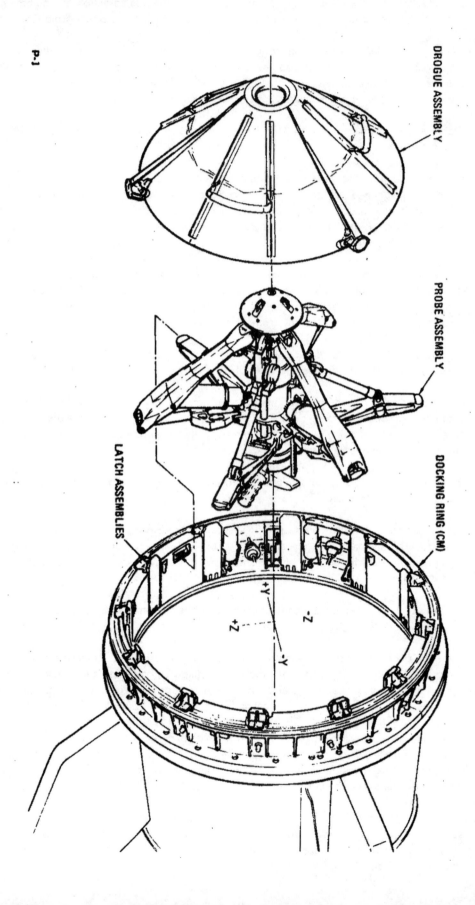

P.1

DROGUE ASSEMBLY

PROBE ASSEMBLY

LATCH ASSEMBLIES

DOCKING RING (CM)

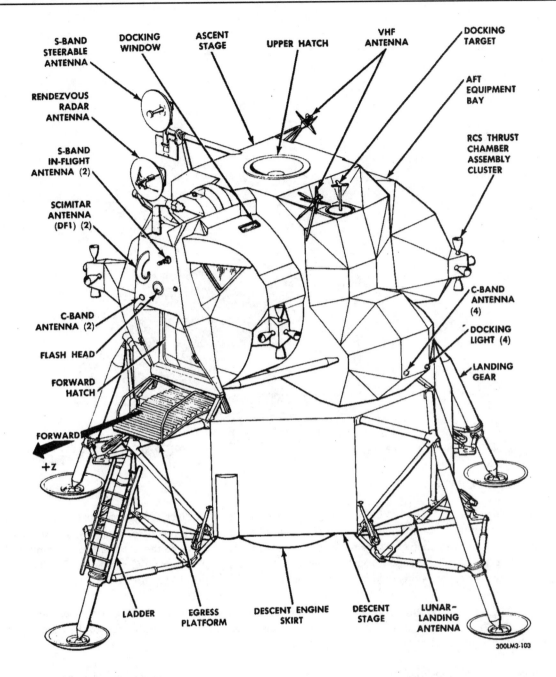

S-BAND STEERABLE ANTENNA
DOCKING WINDOW
ASCENT STAGE
UPPER HATCH
VHF ANTENNA
DOCKING TARGET
RENDEZVOUS RADAR ANTENNA
AFT EQUIPMENT BAY
S-BAND IN-FLIGHT ANTENNA (2)
RCS THRUST CHAMBER ASSEMBLY CLUSTER
SCIMITAR ANTENNA (DF1) (2)
C-BAND ANTENNA (4)
C-BAND ANTENNA (2)
DOCKING LIGHT (4)
FLASH HEAD
LANDING GEAR
FORWARD HATCH
FORWARD
+z
LADDER
EGRESS PLATFORM
DESCENT ENGINE SKIRT
DESCENT STAGE
LUNAR-LANDING ANTENNA

300LM3-103

Each pad is fitted with a lunar-surface sensing probe which signal the crew to shut down the descent engine upon contact with the lunar surface.

LM-3 flown on the Apollo 9 mission will have a launch weight of 32,000 pounds (14,507-8 kg). The weight breakdown is as follows:

Ascent stage, dry	5,071 lbs
Descent stage, dry	4,265 lbs
RCS propellants	605 lbs
DPS propellants	17,944 lbs
APS propellants	4,136 lbs
Total	32,021 lbs

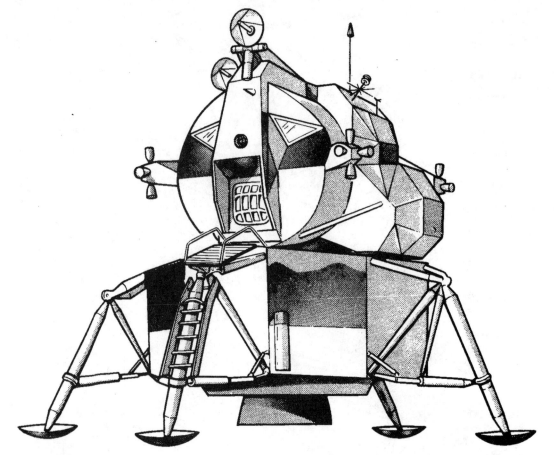

LM-3 Spacecraft Systems

Electrical Power System -- The LM DC electrical system consists of six silver zinc primary batteries - four in the descent stage and two in the ascent stage, each with its own electrical control assembly (ECA). Power feeders from all primary batteries pass through circuit breakers to energize the LM DC buses, from which 28-volt DC power is distributed through circuit breakers to all LM systems. AC power (117v 400Hz) is supplied by two inverters, either of which can supply spacecraft AC load needs to the AC buses.

Environmental Control System -- Consists of the atmosphere revitalization section, oxygen supply and cabin pressure control section, water management, heat transport section and outlets for oxygen and water servicing of the Portable Life Support System (PLSS).

Components of the atmosphere revitalization section are the suit circuit assembly which cools and ventilates the pressure garments, reduces carbon dioxide levels, removes odors and noxious gases and excessive moisture; the cabin recirculation assembly which ventilates and controls cabin atmosphere temperatures; and the steam flex duct which vents to space steam from the suit circuit water evaporator.

The oxygen supply and cabin pressure section supplies gaseous oxygen to the atmosphere revitalization section for maintaining suit and cabin pressure. The descent stage oxygen supply provides descent phase and lunar stay oxygen needs, and the ascent stage oxygen supply provides oxygen needs for the ascent and rendezvous phase.

Water for drinking, cooling, firefighting and food preparation and refilling the PLSS cooling water servicing tank is supplied by the water management section. The water is contained in three nitrogen-pressurized bladder-type tanks, one of 367-pound capacity in the descent stage and two of 47.5-pound capacity in the ascent stage.

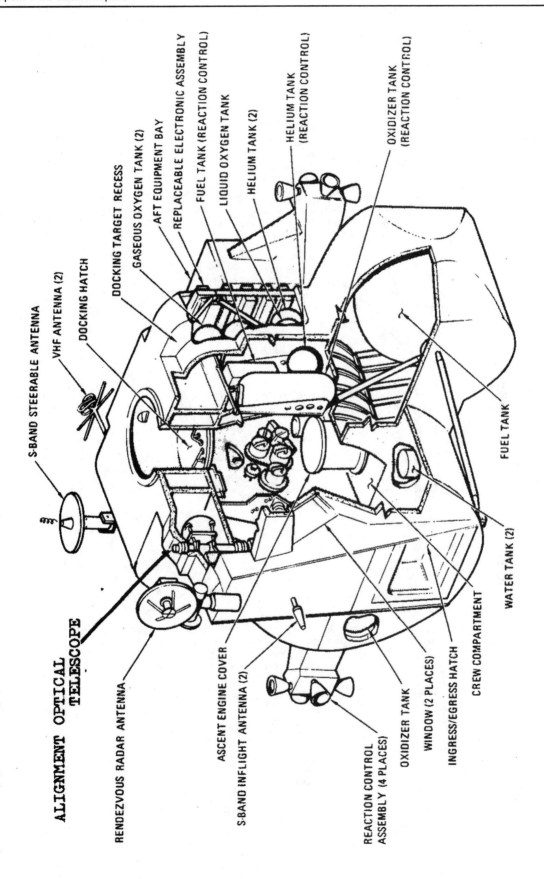

ALIGNMENT OPTICAL TELESCOPE

S-BAND STEERABLE ANTENNA

VHF ANTENNA (2)

DOCKING HATCH

DOCKING TARGET RECESS

GASEOUS OXYGEN TANK (2)

AFT EQUIPMENT BAY

REPLACEABLE ELECTRONIC ASSEMBLY

FUEL TANK (REACTION CONTROL)

LIQUID OXYGEN TANK

HELIUM TANK (2)

HELIUM TANK (REACTION CONTROL)

OXIDIZER TANK (REACTION CONTROL)

FUEL TANK

WATER TANK (2)

CREW COMPARTMENT

INGRESS/EGRESS HATCH

WINDOW (2 PLACES)

OXIDIZER TANK

REACTION CONTROL ASSEMBLY (4 PLACES)

S-BAND INFLIGHT ANTENNA (2)

ASCENT ENGINE COVER

RENDEZVOUS RADAR ANTENNA

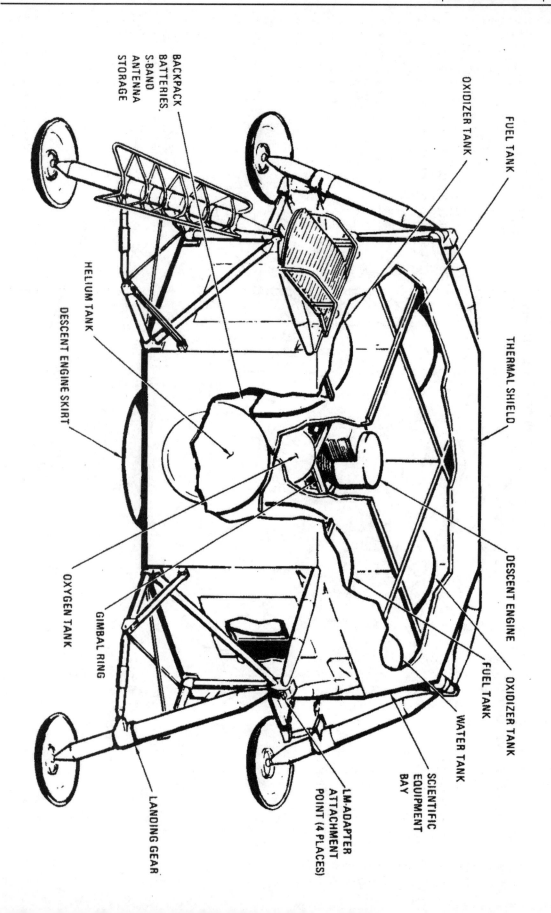

BACKPACK
BATTERIES.
S-BAND
ANTENNA
STORAGE

OXIDIZER TANK

FUEL TANK

HELIUM TANK

DESCENT ENGINE SKIRT

THERMAL SHIELD

DESCENT ENGINE

FUEL TANK

OXIDIZER TANK

OXYGEN TANK

GIMBAL RING

LANDING GEAR

WATER TANK

SCIENTIFIC
EQUIPMENT
BAY

LM-ADAPTER
ATTACHMENT
POINT (4 PLACES)

The heat transport section has primary and secondary water-glycol solution coolant loops. The primary coolant loop circulates water-glycol for temperature control of cabin and suit circuit oxygen and for thermal control of batteries and electronic components mounted on cold plates and rails. If the primary loop becomes inoperative, the secondary loop circulates coolant through the rails and cold plates only. Suit circuit cooling during secondary coolant loop operation is provided by the suit loop water boiler. Waste heat from both loops is vented overboard by water evaporation, or sublimators.

Communication System -- Two S-Band transmitter-receivers, two VHF transmitter-receivers, a UHF command receiver, a signal processing assembly and associated spacecraft antenna make up the LM communications system. The system transmits and receives voice, tracking and ranging data, and transmits telemetry data on 281 measurements and TV signals to the ground. Voice communications between the LM and ground stations is by S-Band, and between the LM and CSM voice is on VHF. In Earth orbital operations such as Apollo 9, VHF voice communications between the LM and the ground are possible. Developmental flight instrumentation (DFI) telemetry data are transmitted to MSFN stations by five VHF transmitters. Two C-Band beacons augment the S-Band system for orbital tracking.

The UHF receiver accepts command signals which are fed to the LM guidance computer for ground updates of maneuvering and navigation programs. The UHF receiver is also used to receive real-time commands which are on LM-3 to arm and fire the ascent propulsion system for the unmanned APS depletion burn. The UHF receiver will be replaced by an S-Band command system on LM-4 and subsequent spacecraft.

The Data Storage Electronics Assembly (DSEA) is a four channel voice recorder with timing signals with a 10-hour recording capacity which will be brought back into the CSM for return to Earth. DSEA recordings cannot be "dumped" to ground stations.

LM antennas are one 26-inch diameter parabolic S-Band steerable antenna, two S-Band in-flight antennas, two VHF in-flight antennas, four C-Band antennas, and two UHF/VHF command/ DFI scimitar antennas.

Guidance, Navigation and Control System -- Comprised of six sections: primary guidance and navigation section (PGNS), abort guidance section (AGS), radar section, control electronics section (CES), and orbital rate drive electronics for Apollo and LM (ORDEAL).

* The PGNS is an inertial system aided by the alignment optical telescope, an inertial measurement unit, and the rendezvous and landing radars. The system provides inertial reference data for computations, produces inertial alignment reference by feeding optical sighting data into the LM guidance computer, displays position and velocity data, computes LM-CSM rendezvous data from radar inputs, controls attitude and thrust to maintain desired LM trajectory and controls descent engine throttling and gimballing.

* The AGS is an independent backup system for the PGNS, having its own inertial sensor and computer.

* The radar section is made up of the rendezvous radar which provides and feeds CSM range and range rate, and line-of sight angles for maneuver computation to the LM guidance computer; the landing radar which provides and feeds altitude and velocity data to the LM guidance computer during lunar landing. On LM-3, the landing radar will be in a self-test mode only. The rendezvous radar has an operating range from 80 feet to 400 nautical miles.

* The CES controls LM attitude and translation about all axes. Also controls by PGNS command the automatic operation of the ascent and descent engines, and the reaction control thrusters. Manual attitude controller and thrust-translation controller commands are also handled by the CES.

* ORDEAL displays on the flight director attitude indicator the computed local vertical in the pitch axis during circular Earth or lunar orbits.

Reaction Control System -- The LM has four RCS engine clusters of four 100-pound (45.4 kg) thrust engines each which use helium-pressurized hypergolic propellants. The oxidizer is nitrogen tetroxide, fuel is Aerozine

50 (50/50 hydrazine and unsymmetrical dimethyl hydrazine). Propellant plumbing, valves and pressurizing components are in two parallel, independent systems, each feeding half the engines in each cluster. Either system is capable of maintaining attitude alone, but if one supply system fails, a propellant crossfeed allows one system to supply all 16 engines. Additionally, interconnect valves permit the RCS system to draw from ascent engine propellant tanks. The engine clusters are mounted on outriggers 90 degrees apart on the ascent stage. The RCS provides small stabilizing impulses during ascent and descent burns, controls LM attitude during maneuvers, and produces thrust for separation and ascent/descent engine tank ullage. The system may be operated in either pulsed or steady state modes.

Descent Propulsion System -- Maximum rated thrust of the descent engine is 9,870 pounds (4,380.9) and is throttleable between 1,050 pounds (476.7 kg) and 6,300 pounds (2,860.2 kg). The engine can be gimbaled six degrees in any direction for offset center of gravity trimming. Propellants are helium pressurized Aerozine 50 and nitrogen tetroxide.

Ascent Propulsion System -- The 3,500 Pound (1,589 kg) thrust ascent engine is not gimbaled and performs at full thrust. The engine remains dormant until after the ascent stage separates from the descent stage. Propellants are the same as are burned in the RCS engines and the descent engine.

Caution and Warning Controls and Displays -- These two systems have the same function aboard the lunar module as they do aboard the command module. (see CSM systems section.)

Tracking and Docking Lights -- A flashing tracking light (once per second, 20 milliseconds duration) on the front face of the lunar module is an aid for contingency CSM-active rendezvous LM rescue. Visibility ranges from 400 nautical miles through the CSM sextant to 130 miles with the naked eye. Five docking lights analogous to aircraft running lights are mounted on the LM for CSM-active rendezvous: two forward yellow lights, aft white light, port red light and starboard green light. All docking lights have about a 1,000-foot visibility.

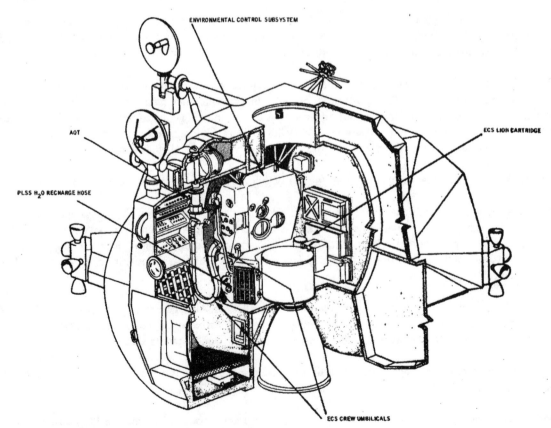

ENVIRONMENTAL CONTROL SUBSYSTEM

AOT

ECS LIOH CARTRIDGE

PLSS H₂0 RECHARGE HOSE

ECS CREW UMBILICALS

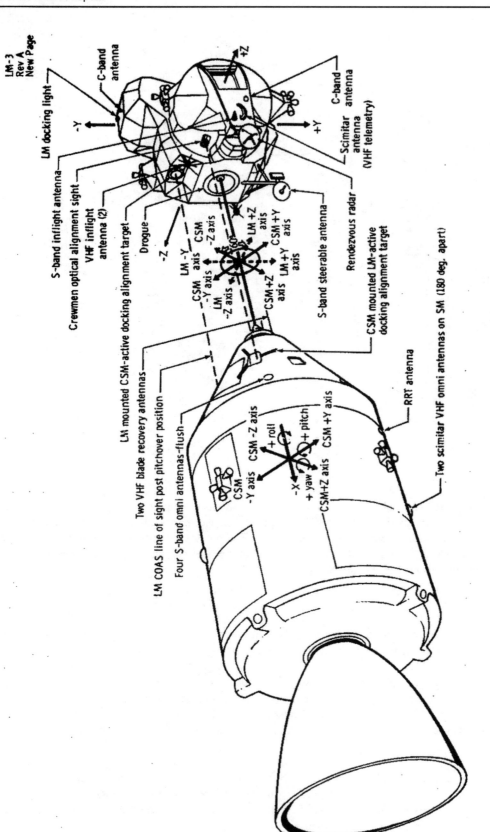

Figure 8-3.2 - LM-CSM antenna locations.

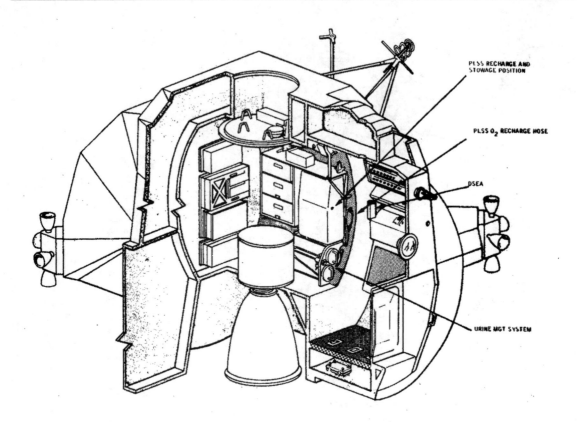

PLSS RECHARGE AND STOWAGE POSITION

PLSS O_2 RECHARGE HOSE

DSEA

URINE MGT SYSTEM

LAUNCH VEHICLE

Saturn V

The Saturn V — 363 feet tall with the Apollo spacecraft in place, generates enough thrust to place a 125 - ton payload into a 105 nm circular Earth orbit or boost a smaller payload to the vicinity of any planet in the solar system. It can boost about 50 tons to lunar orbit. The thrust of the three propulsive stages range from more than 7.7 million pounds for the booster to 230,000 pounds for the third stage at operating altitude. Including the instrument unit, the launch vehicle is 281 feet tall.

First Stage

The first stage (S-IC) was developed jointly by the National Aeronautics and Space Administration's Marshall Space Flight Center, Huntsville, Ala., and the Boeing Co.

The Marshall Center assembled four S-IC stages: a structural test model, a static test version and the first two flight stages. Subsequent flight stages are being assembled by Boeing at the Michoud Assembly Facility in New Orleans. The S-IC stage destined for the Apollo 9 mission was the first flight booster static tested at the NASA-Mississippi Test Facility. That test was made on May 11, 1967. Earlier flight stages were static fired at the NASA-Marshall Center.

The S-IC stage provides first boost of the Saturn V launch vehicle to an altitude of about 37 nautical miles (41.7 statute miles, 67.1 kilometers) and provides acceleration to increase the vehicle's velocity to 9,095 feet per second (2,402 m/sec, 5,385 knots, 6,201 mph). It then separates from the S-II stage and falls to Earth about 361.9 nm (416.9 am, 667.3 km) downrange.

Normal propellant flow rate to the five F-1 engines is 29,522 pounds per second. Four of the engines are mounted on a ring, each 90 degrees from its neighbor. These four are gimbaled to control the rocket's direction of flight. The fifth engine is mounted rigidly in the center.

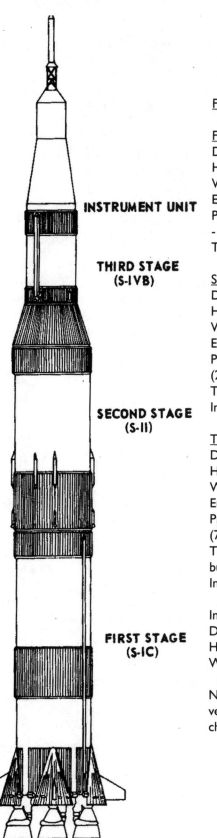

INSTRUMENT UNIT

**THIRD STAGE
(S-IVB)**

**SECOND STAGE
(S-II)**

**FIRST STAGE
(S-IC)**

FACT SHEET, SA-504

First Stage (S-IC)
Diameter ------- 33 feet,
Height -- 138 feet
Weight --------- 5,026,200 lbs. Fueled 295,600 lbs. dry
Engines -------- Five F-1
Propellants ---- Liquid oxygen (347,300 gals.) RP-1 (Kerosene)
- (211,140 gals.)
Thrust --------- 7,700,000 lbs.

Second Stage (S-II)
Diameter--------- 33 feet,
Height -- 81.5 feet
Weight --------- 1,069,033 lbs. Fueled 84,600 lbs. dry
Engines -------- Five J-2
Propellants ---- Liquid oxygen (86,700 gals.) Liquid hydrogen
(281,550 gals.)
Thrust --------- 1,150,000 lbs.
Interstage ----- 10,305 lbs.

Third Stage (S-IVB)
Diameter 21.7 feet,
Height -- 58.3 ft.
Weight --------- 258,038 lbs. Fueled 25,300 lbs, dry
Engines -------- One J-2
Propellants ---- Liquid oxygen (19,600 gals) Liquid hydrogen
(77,675 gals.)
Thrust --------- 232,000 lbs. first burn (211,000 lbs. second
burn)
Interstage----- 8,061 lbs.

Instrument Unit
Diameter 21.7 feet,
Height -- 3 feet
Weight --------- 4,295 lbs.

NOTE: Weights and measures given above are for the nominal
vehicle configuration. The figures may vary slightly due to
changes before or during flight to meet changing conditions.

Second Stage

The second stage (S-II), like the third stage, uses high performance J-2 engines that burn liquid oxygen and liquid hydrogen. The stage's purpose is to provide second stage boost nearly to Earth orbit.

At engine cutoff, the S-II separates from the third stage and, following a ballistic trajectory, plunges into the Atlantic Ocean about 2,412 nm (2,778.6 sm, 4,468 km) downrange from Cape Kennedy.

Five J-2 engines power the S-II. The four outer engines are equally spaced on a 17.5-foot diameter circle. These four engines may be gimbaled through a plus or minus seven-degree square pattern for thrust vector control. Like the first stage, the center (number 5) engine is mounted on the stage centerline and is fixed.

The S-II carries the rocket to an altitude of 103 nm (118.7 sm, 190.9 km) and a distance of some 835 nm (961.9 sm. 1,548 km) downrange. Before burnout, the vehicle will be moving at a speed of 23,000 fps or 13,642 knots (15,708 mph, 25,291 kph, 6,619 m/sec). The J-2 engines will burn six minutes eleven seconds during this powered phase.

The Space Division of North American Rockwell Corp. builds the S-II at Seal Beach, Calif. The cylindrical vehicle is made up of the forward skirt (to which the third stage attaches), the liquid hydrogen tank, the liquid oxygen tank, the thrust structure (on which the engines are mounted) and an interstage section (to which the first stage connects). The propellant tanks are separated by an insulated common bulkhead.

The S-II was static tested by North American Rockwell at the NASA-Mississippi Test Facility on Feb. 10, 1968. This Apollo 9 flight stage was shipped to the test site via the Panama Canal for the test firing.

Third Stage

The third stage (S-IVB) was developed by the McDonnell Douglas Astronautics Co. at Huntington Beach, Calif.. At Sacramento Calif. the stage passed a static firing test on Aug. 26, 1966, as part of the preparation for the Apollo 9 mission. The stage was flown directly to the NASA-Kennedy Space Center.

Measuring 58 feet 4 inches long and 21 feet 8 inches in diameter, the S-IVB weighs 25,300 pounds dry. At first ignition it weighs 259,377 pounds. The interstage section weighs an additional 8,081 pounds. The stage's J-2 engine burns liquid oxygen and liquid hydrogen.

The stage, with its single engine, provides propulsion three times during the Apollo 9 mission. The first burn occurs immediately after separation from the S-II. It will last long enough (112 seconds) to insert the vehicle and spacecraft into Earth parking orbit. The second burn, which begins after separation from the spacecraft, will place the stage and instrument unit into a high apogee elliptical orbit. The third burn will drive the stage into solar orbit.

The fuel tanks contain 77,675 gallons of liquid hydrogen and 19,600 gallons of liquid oxygen at first ignition, totaling 230,790 pounds of propellants. Insulation between the two tanks is necessary because the liquid oxygen, at about 293 degrees below zero F, is warm enough, relatively, to heat the liquid hydrogen, at 423 degrees below zero F, rapidly and cause it to change into a gas.

The first re-ignition burn is for 62 seconds and the second re-ignition burn is planned to last 4 minutes 2 seconds. Both re-ignitions will be inhibited with inhibit removal to be by ground command only after separation of the spacecraft to a safe distance.

Instrument Unit

The Instrument Unit (IU) is a cylinder three feet high and 21 feet 8 inches in diameter. It weighs 4,295 pounds and contains the guidance, navigation and control equipment which will steer the vehicle through its Earth orbits and into the final escape orbit maneuver.

The IU also contains telemetry, communications, tracking and crew safety systems, along with its own supporting electrical power and environmental control systems.

Components making up the "brain" of the Saturn V are mounted on cooling panels fastened to the inside surface of the instrument unit skin. The "cold plates" are part of a system that removes heat by circulating cooled fluid through a heat exchanger that evaporates water from a separate supply into the vacuum of space.

The six major systems of the instrument unit are structural, thermal control, guidance and control, measuring and telemetry, radio frequency, and electrical.

The instrument unit provides navigation, guidance and control of the vehicle measurement of vehicle performance and environment; data transmission with ground stations; radio tracking of the vehicle; checkout and monitoring of vehicle functions; initiation of stage functional sequencing; detection of emergency situations; generation and network, distribution of electric power for system operation; and preflight checkout and launch and flight operations.

A path-adaptive guidance scheme is used in the Saturn V instrument unit. A programmed trajectory is used in the initial launch phase with guidance beginning only after the vehicle has left the atmosphere. This is to prevent movements that might cause the vehicle to break apart while attempting to compensate for winds, jet streams and gusts encountered in the atmosphere.

If such air currents displace the vehicle from the optimum trajectory in climb, the vehicle derives a new trajectory. Calculations are made about once each second throughout the flight. The launch vehicle digital computer and launch vehicle data adapter perform the navigation and guidance computations.

The ST- 124M inertial platform -- the heart of the navigation, guidance and control system -- provides space-fixed reference coordinates and measures acceleration along the three mutually perpendicular axes of the coordinate system.

International Business Machines Corp., is prime contractor for the instrument unit and is the supplier of the guidance signal processor and guidance computer. Major suppliers of instrument unit components are: Electronic Communications Inc., control computer; Bendix Corp., ST-124M inertial platform; and IBM Federal Systems Division, launch vehicle digital computer and launch vehicle data adapter,

Propulsion

The 41 rocket engines of the Saturn V have thrust ratings ranging from 72 pounds to more than 1.5 million pounds. Some engines burn liquid propellants, others use solids.

The five F-1 engines in the first stage burn RP-1 (kerosene) and liquid oxygen. Each engine in the first stage develops an average of 1,544,000 pounds of thrust at liftoff, building up to an average of 1,833,900 pounds before cutoff. The cluster of five engines gives the first stage a thrust range from 7.72 million pounds at liftoff to 9,169,560 pounds just before center engine cutoff.

The F-1 engine weighs almost 10 tons, is more than 18 feet high and has a nozzle-exit diameter of nearly 14 feet. The F-1 undergoes static testing for an average 650 seconds in qualifying for the 150-second run during the Saturn V first stage booster phase. This run period, 800 seconds, is still far less than the 2,200 seconds of the engine guarantee period. The engine consumes almost three tons of propellants per second.

The first stage of the Saturn V for this mission has eight other rocket motors. These are the solid-fuel retro-rockets which will slow and separate the stage from the second stage. Each rocket produces a thrust of 87,900 pounds for 0.6 second.

The main propulsion for the second stage is a cluster of five J-2 engines burning liquid hydrogen and liquid oxygen. Each engine develops a mean thrust of more than 205,000 pounds at 5.0:1 mixture ratio (variable from 193,000 to 230,000 in phases of flight), giving the stage a total mean thrust of more than a million pounds. Designed to operate in the hard vacuum of space, the 3,500 pound J-2 is more efficient than the F-1 because it burns the high-energy fuel hydrogen.

The second stage also has four 21,000 pound thrust solid fuel rocket engines. These are the ullage rockets mounted on the S-IC/S-II interstage section. These rockets fire to settle liquid propellant in the bottom of the main tanks and help attain a "clean" separation from the first stage, they remain with the interstage when it drops away at second plane separation. Four retrorockets are located in the S-IVB aft interstage (which never separates from the S-II) to separate the S-II from the S-IVB prior to S-IVB ignition.

Eleven rocket engines perform various functions on the third stage. A single J-2 provides the main propulsive force; there are two jettisonable main ullage rockets and eight smaller engines in the two auxiliary propulsion system modules.

Launch Vehicle Instrumentation and Communication

A total of 2,159 measurements will be taken in flight on the Saturn V launch vehicle: 666 on the first stage, 975 on the second stage, 296 on the third stage and 222 on the instrument unit,

The Saturn V will have 16 telemetry systems: six on the first stage, six on the second stage, one on the third stage and three on the instrument unit. A radar tracking system will be on the first stage, and a C-Band system and command system on the instrument unit. Each powered stage will have a range safety system as on previous flights. There will be no film or television cameras on or in any of the stages of the Saturn 504 launch vehicle.

Vehicle Weights During Flight

Event	Vehicle Weight
Ignition	6,486,915 pounds
Liftoff	6,400,648 pounds
Mach 1	4,487,938 pounds
Max. Q	4,015,350 pounds
CECO	2,443,281 pounds
OECO	1,831,574 pounds
S-IC/S-II Separation	1,452,887 pounds
S-II Ignition	1,452,277 pounds
Interstage Jettison	1,379,282 pounds
LET Jettison	1,354,780 pounds
S-II Cutoff	461,636 pounds
S-II/S-IVB Separation	357,177 pounds
S-IVB Ignition	357,086 pounds
S-IVB First Cutoff	297,166 pounds
Parking Orbit Injection	297,009 pounds
Spacecraft First Separation	232,731 pounds (Only S-IVB and LM)
Spacecraft Docking	291,572 pounds (S-IVB and complete spacecraft)
Spacecraft Second Separation	199,725 pounds (only S-IVB stage)
S-IVB First Re-ignition	199,346 pounds
S-IVB Second Cutoff	170,344 pounds
Intermediate Orbit Injection	170,197 pounds
S-IVB Second Re-ignition	169,383 pounds
S-IVB Third Cutoff	54,440 pounds
Escape Orbit Injection	54,300 pounds
End LOX Dump	35,231 pounds
End LH2 Dump	31,400 pounds

S-IVB Restarts

The third stage (S-IVB) of the Saturn V rocket for the Apollo 9 mission will burn a total of three times in space, the last two burns unmanned for engineering evaluation of stage capability. It has never burned more than twice in space before.

Engineers also want to check out the primary and backup propellant tank pressurization systems and prove that in case the primary system fails the backup system will pressurize the tanks sufficiently for restart.

Also planned for the second restart is an extended fuel lead for chilldown of the J-2 engine using liquid hydrogen fuel flowing through the engine to chill it to the desired temperature prior to restart.

All these events -- second restart, checkout of the backup pressurization system and extended fuel lead chilldown will not affect the primary mission of Apollo 9. The spacecraft will have been separated and will be safe in a different orbit from that of the spacecraft before the events begin.

Previous flights of the S-IVB stage have not required a second restart, and none of the flights currently planned have a specific need for the third burn.

The second restart during the Apollo 9 mission will come 80 minutes after second engine burn cutoff to demonstrate a requirement that the stage has the capability to restart in space after being shut down for only 80 minutes. That capability has not yet been proven because flights to date have not required as little as 80 minutes of coastings.

The first restart is scheduled for four hours 45 minutes and 41 seconds after launch, or about four and one-half hours after first burn cutoff. The need for a second restart may occur on future non-lunar flights. Also, the need to launch on certain days could create a need for the 80-minute restart capability.

The primary pressurization system of the propellant tanks for S-IVB restart uses a helium heater. In this system, nine helium storage spheres in the liquid hydrogen tank contain gaseous helium charged to about 3,000 psi. This helium is passed through the heater which heats and expands the gas before it enters the propellant tanks. The heater operates on hydrogen and oxygen gas from the main propellant tanks.

The backup system consists of five ambient helium spheres mounted on the stage thrust structure. This system, controlled by the fuel repressurization control module, can repressurize the tanks in case the primary system fails.

The first restart will use the primary system. If that system fails, the backup system will be used. The backup system will be used for the second restart.

The primary reason for the extended fuel lead chilldown test is to demonstrate a contingency plan in case the chilldown pumps fail. In the extended chilldown event, a ground command will cut off the pumps to put the stage in a simulated failure condition. Another ground command will start liquid hydrogen flowing through the engine.

This is a slower method, but enough time has been allotted for the hydrogen to chill the engine and create about the same conditions as would be created by the primary system.

The two unmanned restarts in this mission are at the very limits of the design requirements for the S-IVB and the J-2 engine. For this reason, the probability of restart is not as great as in the nominal lunar missions.

Two requirements are for the engine to restart four and one-half hours after first burn cutoff, and for a total stage lifetime of six and one-half hours. The first restart on this mission will be four and one-half hours after first burn cutoff, and second restart will occur six hours and six minutes after liftoff, both events scheduled near the design limits.

Differences in Apollo 8 and Apollo 9 Launch Vehicles

Two modifications resulting from problems encountered during the second Saturn V flight were incorporated and proven successful on the third Saturn V mission. The new helium prevalve cavity pressurization system will again be flown on the S-IC stage of Apollo 9. Also, new augmented spark igniter lines which flew on the engines of the two upper stages of Apollo 8 will again be used on Apollo 9.

The major S-IC stage differences between Apollo 8 and Apollo 9 are:

1. Dry weight was reduced from 304,000 pounds to 295,600 pounds.

2. Weight at ground ignition increased from 4,800,000 pounds to 5,026,200 pounds.

3. Instrumentation measurements were reduced from 891 to 666.

4. Camera instrumentation electrical power system is not installed on S-IC-4.

5. S-IC-4 carries neither a TV camera system nor a film camera system.

The Saturn V will fly a somewhat lighter and slightly more powerful second stage beginning with Apollo 9.

The changes are:

1. Nominal vacuum thrust for J-2 engines was increased from 225,000 pounds each to 230,000 pounds each. This changed the second stage thrust from a total of 1,125,000 pounds to 1,150,000 pounds.

2. The approximate empty S-II stage weight has been reduced from 88,000 to 84,600 pounds. The S-IC/S-II interstage weight was reduced from 11,800 to 11,664 pounds.

3. Approximate stage gross liftoff weight was increased from 1,035,000 pounds to 1,069,114 pounds.

4. S-II instrumentation system was changed from research and development to a combination of research and development and operational.

5. Instrumentation measurements were decreased from 1,226 to 975.

Major differences between the S-IVB stage used on Apollo 8 and the one on Apollo 9 are:

1. S-IVB dry stage weight decreased from 26,421 pounds to 25,300 pounds. This does not include the 8,081-pound interstage section.

2. S-IVB gross stage weight at liftoff decreased from 263,204 pounds to 259,337 pounds.

3. Stage measurements evolved from research and development to operational status.

4. Instrumentation measurements were reduced from 342 to 296.

Major instrument unit differences between Apollo 8 and Apollo 9 include deletions of a rate gyro timer, thermal probe, a measuring distributor, a tape recorder, two radio frequency assemblies, a source follower, a battery and six measuring racks. Instrumentation measurements were reduced from 339 to 222.

Launch Vehicle Sequence of Events

(Note: Information presented in this press kit is based upon a nominal mission. Plans may be altered prior to or during flight to meet changing conditions.)

Launch

The first stage of the Saturn V will carry the launch vehicle and Apollo spacecraft to an altitude of 36.2 nautical miles (41.7 sm, 67.1 km) and 50 nautical miles (57.8 sm. 93 km) downrange, building up speed to 9,095.2 feet per second (2,402 m/sec, 5,385.3 knots, 6,201.1 mph) in two minutes 31 seconds of powered flight. After separation from the second stage, the first stage will continue on a ballistic trajectory ending in the Atlantic Ocean some 361.9 nautical miles (416.9 sm, 670.9 km) downrange from Cape Kennedy (latitude 30.27 degrees north and longitude 73.9 degrees west) about nine minutes after liftoff.

Second Stage

The second stage, with engines running six minutes and 11 seconds, will propel the vehicle to an altitude of about 103 nautical miles (118.7 sm. 190.9 km) some 835 nautical miles (961.7 sm, 1,547 km) downrange, building up to 23,040.3 feet per second (6,619 m/sec, 13,642.1 knots, 15,708.9 mph) space fixed velocity. The spent second stage will land in the Atlantic Ocean about 20 minutes after liftoff some 2,410.3 nautical miles (2,776.7 sm, 4,468.4 km) from the launch site, at latitude 31.46 degrees north and longitude 34.06 degrees west.

Third Stage First Burn

The third stage, in its 112-second initial burn, will place itself and the Apollo spacecraft into a circular orbit 103 nautical miles (119 sm, 191 km) above the Earth. Its inclination will be 32.5 degrees and the orbital period about 88 minutes. Apollo 9 will enter orbit at about 56.66 degrees west longitude and 32.57 degrees north latitude.

Parking Orbit

The Saturn V third stage will be checked out in Earth parking orbit in preparation for the second S-IVB burn. During the second revolution, the Command/Service Module (CSM) will separate from the third stage. The spacecraft LM adapter (SLA) panels will be jettisoned and the CSM will turn around and dock with the lunar module (LM) while the LM is still attached to the S-IVB/IU. This, maneuver is scheduled to require 14 minutes. About an hour and a quarter after the docking, LM ejection is to occur. A three second SM RCS burn will provide a safe separation distance at S-IVB re-ignition. All S-IVB re-ignitions are nominally inhibited. The inhibit is removed by ground command after the CSM/LM is determined to be a safe distance away.

Third Stage Second Burn

Boost of the unmanned S-IVB stage from Earth parking orbit to an intermediate orbit occurs during the third revolution shortly after the stage comes within range of Cape Kennedy (4 hours 44 minutes 42 seconds after liftoff). The second burn, lasting 62 seconds, will put the S-IVB into an elliptical orbit with an apogee of 3,052 km (1,646.3 nm, 1,896.5 sm) and a perigee of 196 km (105.7 nm, 121.8 sm). The stage will remain in this orbit about one-half revolution.

Third Stage Third Burn

The third stage is re-ignited at 6 hours 6 minutes 4 seconds after liftoff for a burn lasting 4 minutes 2 seconds. This will place the stage and instrument unit into the escape orbit. Ninety seconds after cutoff, LOX dump begins and lasts for 11 minutes 10 seconds, followed 10 seconds later by the dump of the liquid hydrogen, an exercise of 18 minutes 15 seconds duration. Total weight of the stage and instrument unit placed into solar orbit will be about 31,400 pounds.

Launch Vehicle Key Event

Time		Altitude				Velocity			
Hrs Min Sec	Event	Meters	Feet	N. Mi.	S. Mi.	M/Sec.	F/Sec.	MPH	Knots
00 00 00	First Motion	60	198	.033	.038	0.00	0.00	0.0	0.00
00 00 12	Tilt Initiation	225	737	.12	.14	410.00	1345.4	279.5	242.8
00 01 21	Maximum Dynamic Pressure	13311	43671	7.2	8.3	816.00	2677.00	1825.2	1585.1
00 02 14	Center Engine Cutoff	44981	147576	24.3	27.9	2005. 6	6580.2	4486.4	3896. 1
00 02 39	Outboard Engine Cutoff	67132	220248	36.2	41.7	2772.2	9095.2	6201.1	5385.3
00 02 40	S-IC/S-II Separation	67848	222599	36.6	42.2	2781.9	9127.0	6222.8	5404.1
00 02 42	S-II Ignition	69354	227541	37.4	43.1	2777.7	9113.3	6213.4	5396.0
00 03 10	S-II Aft Interstage Jettison	92816	304514	50.1	57.7	2909.6	9546.0	6508.5	5652.2
00 03 15	LES Jettison	97075	318487	52.4	60.3	2942.6	9654.2	6582.2	5716.3
00 03 21	Initiate IGM	100747	330534	54.3	62.6	2973.3	9754.8	6650.8	5775.8
00 08 53	S-II Cutoff	190956	626497	103.0	118.7	7022.7	23040.3	15708.9	13642.2
00 08 54	S-II/S-IVB Separation	190998	626633	103.0	118.7	7027.3	23055.5	15719.2	13651.2
00 08 57	S-IVB Ignition	191137	627089	103.1	118.8	7027.6	23056.5	15719.9	13651.8
00 10 49	S-IVB First Cutoff	191385	627904	103.2	118.9	7791.0	25561.1	15427.6	15134.7
00 10 59	Parking Orbit Insertion	191398	627947	103.2	118.9	7793.0	25567.7	17432.1	15138.6
02 43 00	Spacecraft Separation	195960	642913	105.7	121.8	7790.8	25560.4	17427.1	15134.3
02 53 43	Spacecraft Docking	198552	651416	107.1	123.4	7787.9	25551.0	17420.7	15128.7
04 08 57	Spacecraft Final Separation	194366	637682	104.8	120.8	7792.3	25565.2	17430.4	15137.2
04 45 41	S-IVB Re-ignition	199349	655671	107.8	124.2	7789.6	25556.4	17424.4	15131.9
04 46 43	S-IVB Second Cutoff	200175	656744	108.0	124.4	8449.0	27719.8	18899.4	16412.9
04 46 53	Intermediate Orbit Insertion	200720	658532	108.3	124.7	8451.5	27728.2	18905.1	16417.9
06 07 04	S-IVB Re-ignition	2410165	7907364	1300.0	1497.6	6400.1	20997.8	14316.3	12432.8
06 11 05	S-IVB Third Cutoff	2242364	7356839	1209.5	1393.3	11230.8	36846.4	25121.9	21816.8
06 11 15	Escape Orbit Insertion	2244556	7364028	1210.7	1394.7	11239.3	36874.2	25140.8	21833.2
06 12 36	Start LOX Dump	2295412	7530878	1238.1	1426.3	11215.2	36795.3	25087.0	21786.5
06 23 46	LOX Dump Cutoff	4662286	15296215	2514.8	2897.0	10372.6	34030.8	23202.2	20149.6
06 23 56	Start LH2 Dump	4716809	15475094	2544.2	2930.9	10355.5	33974.6	23163.9	20116.4
06 42 11	LH2 Dump Cutoff	11929321	39138193	6434.5	7412.5	8911.5	29237.2	19933.9	17311.3

LAUNCH FACILITIES

Kennedy Space Center-Launch Complex

NASA's John F. Kennedy Space Center performs preflight checkout, test and launch of the Apollo 9 space vehicle. A government-industry team of about 550 will conduct the final countdown from Firing Room 2 of the Launch Control Center (LCC).

The firing room team is backed up by more than 5,000 persons who are directly involved in launch operations at KSC from the time the vehicle and spacecraft stages arrive at the center until the launch is completed.

Initial checkout of the Apollo spacecraft is conducted in work stands and in the altitude chambers in the Manned Spacecraft Operations Building (MSOB) at Kennedy Space Center. After completion of checkout there, the assembled spacecraft is taken to the Vehicle Assembly Building (VAB) and mated with the launch vehicle. There the first integrated spacecraft and launch vehicle tests are conducted. The assembled space vehicle is then rolled out to the launch pad for final preparations and countdown to launch.

In August 1968 a decision was made not to fly Lunar Module 3 on Apollo 8 as had been originally planned. Checkout continued with the remainder of the Apollo 8 space vehicle and Lunar Module 3 was integrated with the test schedule for Apollo 9.

LM-3 arrived at KSC in June 1968 and at the time the decision was made to fly it on Apollo 9 it had just completed systems tests in the MSOB. It was moved to the vacuum chamber later in August and four manned altitude chamber tests were conducted in September. During these tests, the chamber was pumped down to simulate altitudes in excess of 200,000 feet (60,960 meters) and the lunar module and crew systems were thoroughly checked. The prime crew of Spacecraft Commander James McDivitt and Lunar Module Pilot Russell Schweickart participated in two of the runs and the backup crew of Charles Conrad and Alan Bean participated in the other two runs.

The Apollo 9 command/service module arrived at KSC in October and after receiving inspection in the MSOB, a docking test was conducted with the LM. Two manned altitude chamber runs were made in November with the prime crew participating in one and the backups in the other.

In December the LM was mated to the spacecraft lunar module adapter (SLA), the command/service module was mated to the SLA, and the assembled spacecraft was moved to the VAB where it was erected on the Saturn V launch vehicle (SA 504).

The Apollo 9 launch vehicle had been assembled on its mobile launcher in the VAB in early October. Tests were conducted on individual systems on each of the stages and on the overall vehicle before the spacecraft was mated.

After spacecraft erection, the spacecraft and launch vehicle were electrically mated and the first overall test (plugs-in) of the space vehicle was conducted. In accordance with the philosophy of accomplishing as much of the checkout as possible in the VAB, the overall test was conducted before the space vehicle was moved to the launch pad.

The plugs-in test verified the compatibility of the space vehicle systems, ground support equipment and off-site support facilities by demonstrating the ability of the systems to proceed through a simulated countdown, launch and flight. During the simulated flight portion of the test, the systems were required to respond to both emergency and normal flight conditions.

The move to the launch pad was conducted Jan. 3. Because minimum pad damage was incurred from the launch of Apollo 8, it was possible to refurbish the pad and roll out Apollo 9 less than two weeks later. The 3½-mile (5-6 km) trip to the pad aboard the transporter was completed in about eight hours.

The space vehicle Flight Readiness Test was conducted late in January. Both the prime and backup crews participate in portions of the FRT, which is a final overall test of the space vehicle systems and ground support equipment when all systems are as near as possible to a launch configuration.

After hypergolic fuels were loaded aboard the space vehicle, the launch vehicle first stage fuel (RP-1) was brought aboard and the final major test of the space vehicle began. This was the countdown demonstration test (CDDT), a dress rehearsal for the final countdown to launch. The CDDT for Apollo 9 was divided into a "wet" and a "dry" portion. During the first, or "wet" portion, the entire countdown, including propellant loading, was carried out down to T-8.9 seconds. The astronaut crews did not participate in the wet CDDT. At the completion of the wet CDDT, the cryogenic propellants (liquid oxygen and liquid hydrogen) were off-loaded, and the final portion of the countdown was re-run, this time simulating the fueling and with the prime astronaut crew participating as they will on launch day.

During the assembly and checkout operations for Apollo 9, launch crews at Kennedy Space Center completed the preparation and launch of Apollo 7 (Oct. 11, 1968) and Apollo 8 (Dec. 21, 1968) and began the assembly and checkout operations for Apollo 10 and Apollo 11 to be launched later this year.

Because of the complexity involved in the checkout of the 363-foot-tall (110.6 meters) Apollo/Saturn V configuration, the launch teams make use of extensive automation in their checkout. Automation is one of the major differences in checkout used on Apollo compared to the procedures used in the Mercury and Gemini programs.

RCA 110A computers, data display equipment and digital data techniques are used throughout the automatic checkout from the time the launch vehicle is erected in the VAB through liftoff. A similar, but separate computer operation called ACE (Acceptance Checkout Equipment) is used to verify the flight readiness of the spacecraft. Spacecraft checkout is controlled from separate firing rooms located in the Manned Spacecraft Operations Building.

KSC Launch Complex 39

Launch Complex 39 facilities at the Kennedy Space Center were planned and built specifically for the Saturn V program, the space vehicle that will be used to carry astronauts to the Moon.

Complex 39 introduced the mobile concept of launch operations, a departure from the fixed launch pad techniques used previously at Cape Kennedy and other launch sites. Since the early 1950's when the first ballistic missiles were launched, the fixed launch concept had been used on NASA missions. This method called for assembly, checkout and launch of a rocket at one site - the launch pad. In addition to tying up the pad, this method also often left the flight equipment exposed to the outside influences of the weather for extended periods.

Using the mobile concept, the space vehicle is thoroughly checked in an enclosed building before it is moved to the launch pad for final preparations. This affords greater protection, a more systematic checkout process using computer techniques, and a high launch rate for the future, since the pad time is minimal.

Saturn V stages are shipped to the Kennedy Space Center by ocean-going vessels and specially designed aircraft, such as the Guppy. Apollo spacecraft modules are transported by air. The spacecraft components are first taken to the Manned Spacecraft Operations Building for preliminary checkout. The Saturn V stages are brought immediately to the Vehicle Assembly Building after arrival at the nearby turning basin.

Apollo 9 is the fourth Saturn V to be launched from Pad A, Complex 39. The historic first launch of the Saturn V, designated Apollo 4, took place Nov. 9, 1967 after a perfect countdown and on-time liftoff at 7 a.m. EST. The second Saturn V mission -- Apollo 6 -- was conducted last April 4. The third Saturn V mission, Apollo 8, was conducted last Dec. 21-27,

The major components of Complex 39 include: (1) the Vehicle Assembly Building VAB) where the Apollo 9 was assembled and prepared; (2) the Launch Control Center, where the launch team conducts the preliminary checkout and countdown; (3) the mobile launcher, upon which the Apollo 9 was erected for checkout and from where it will be launched; (4) the mobile service structure, which provides external access to the space vehicle at the pad; (5) the transporter, which carries the space vehicle and mobile launcher, as well as the mobile service structure to the pad; (6) the crawlerway over which the space vehicle travels from the VAB to the launch pad, and (7) the launch pad itself.

The Vehicle Assembly Building

The Vehicle Assembly Building is the heart of Launch Complex 39. Covering eight acres, it is where the 363-foot tall space vehicle is assembled and tested.

The VAB contains 129,482,000 cubic feet of space. It is 716 feet long, and 518 feet wide and it covers 343,500 square feet of floor space.

The foundation of the VAB rests on 4,225 steel pilings, each 16 inches in diameter, driven from 150 to 170 feet to bedrock. If placed end to end, these piles would extend a distance of 123 miles. The skeletal structure of the building contains approximately 60,000 tons of structural steel. The exterior is covered by more than a million square feet of insulated aluminum siding.

The building is divided into a high bay area 525 feet high and a low bay area 210 feet high, with both areas serviced by a transfer aisle for movement of vehicle stages.

The low bay work area, approximately 442 feet wide and 274 feet long, contains eight stage-preparation and checkout cells. These cells are equipped with systems to simulate stage interface and operation with other stages and the instrument unit of the Saturn V launch vehicle.

After the Apollo 9 launch vehicle upper stages arrived at the Kennedy Space Center, they were moved to the low bay of the VAB. Here, the second and third stages underwent acceptance and checkout testing prior to mating with the S-IC first stage atop mobile launcher No. 2 in the high bay area.

The high bay provides the facilities for assembly and checkout of both the launch vehicle and spacecraft. It contains four separate bays for vertical assembly and checkout. At present, three bays are equipped, and the fourth will be reserved for possible changes in vehicle configuration.

Work platforms, some as high as three-story buildings, in the high bays provide access by surrounding the launch vehicle at varying levels. Each high bay has five platforms. Each platform consists of two bi-parting sections that move in from opposite sides and mate, providing a 360-degree access to the section of the space vehicle being checked.

A 10,000-ton-capacity air conditioning system, sufficient to cool about 3,000 homes, helps to control the environment within the entire office, laboratory, and workshop complex located inside the low bay area of the VAB. Air conditioning is also fed to individual platform levels located around the vehicle.

There are 141 lifting devices in the VAB, ranging from one-ton hoists to two 250-ton high-lift bridge cranes.

The mobile launchers, carried by transporter vehicles, move in and out of the VAB through four doors in the high bay area, one in each of the bays. Each door is shaped like an inverted T. They are 152 feet wide and 114 feet high at the base, narrowing to 76 feet in width. Total door height is 456 feet.

The lower section of each door is of the aircraft hangar type that slides horizontally on tracks. Above this are seven telescoping vertical lift panels stacked one above the other, each 50 feet high and driven by an individual motor. Each panel slides over the next to create an opening large enough to permit passage of the Mobile Launcher.

The Launch Control Center

Adjacent to the VAB is the Launch Control Center (LCC). This four-story structure is a radical departure from the dome-shaped blockhouses at other launch sites.

The electronic "brain" of Launch complex 39, the LCC was used for checkout and test operations while Apollo 9 was being assembled inside the VAB. The LCC contains display, monitoring, and control equipment used for both checkout and launch operations.

The building has telemeter checkout stations on its second floor, and four firing rooms, one for each high bay of the VAB, on its third floor. Three firing rooms will contain identical sets of control and monitoring equipment, so that launch of a vehicle and checkout of others may take place simultaneously. A ground computer facility is associated with each firing room.

The high speed computer data link is provided between the LCC and the mobile launcher for checkout of the launch vehicle. This link can be connected to the mobile launcher at either the VAB or at the pad.

The three equipped firing rooms have some 450 consoles which contain controls and displays required for the checkout process. The digital data links connecting with the high bay areas of the VAB and the launch pads carry vast amounts of data required during checkout and launch.

There are 15 display systems in each LCC firing room, with each system capable of providing digital information instantaneously.

Sixty television cameras are positioned around the Apollo/Saturn V transmitting pictures on 10 modulated channels. The LCC firing room also contains 112 operational intercommunication channels used by the crews in the checkout and launch countdown.

Mobile Launcher

The mobile launcher is a transportable launch base and umbilical tower for the space vehicle. Three launchers are used at Complex 39.

The launcher base is a two-story steel structure, 25 feet high, 160 feet long, and 135 feet wide. It is positioned on six steel pedestals 22 feet high when in the VAB or at the launch pad. At the launch pad, in addition to the six steel pedestals, four extendable columns also are used to stiffen the mobile launcher against rebound loads, if the engine cuts off.

The umbilical tower, extending 398 feet above the launch platform, is mounted on one end of the launcher base. A hammerhead crane at the top has a hook height of 376 feet above the deck with a traverse radius of 85 feet from the center of the tower.

The 12-million-pound mobile launcher stands 445 feet high when resting on its pedestals. The base, covering about half an acre, is a compartmented structure built of 25-foot steel girders.

The launch vehicle sits over a 45-foot-square opening which allows an outlet for engine exhausts into a trench containing a flame deflector. This opening is lined with a replaceable steel blast shield, independent of the structure, and will be cooled by a water curtain initiated two seconds after liftoff.

There are nine hydraulically operated service arms on the umbilical tower. These service arms support lines for the vehicle umbilical systems and provide access for personnel to the stages as well as the astronaut crew to the spacecraft.

On Apollo 9, one of the service arms is retracted early in the count. The Apollo spacecraft access arm is partially retracted at T-43 minutes. A third service arm is released at T-30 seconds, and a fourth at about T-6 seconds. The remaining five arms are set to swing back at vehicle first motion after T-0.

The service arms are equipped with a backup retraction system in case the primary mode fails.

The Apollo access arm (service arm No. 9), located at the 320-foot level above the launcher base, provides access to the spacecraft cabin for the closeout team and astronaut crews.

'The flight crew will board the spacecraft starting at about T-2 hours, 40 minutes in the count. The access arm will be moved to a parked position, 12 degrees from the spacecraft, at about T-43 minutes.

This is a distance of about three feet, which permits a rapid reconnection of the arm to the spacecraft in the event of an emergency condition. The arm is fully retracted at the T-5 minute mark in the count.

The Apollo 9 vehicle is secured to the mobile launcher by four combination support and hold-down arms mounted on the launcher deck. The hold-down arms are cast in one piece, about 6 X 9 feet at the base and 10 feet tall, weighing more than 20 tons. Damper struts secure the vehicle near its top.

After the engines ignite, the arms hold Apollo 9 for about six seconds until the engines build up to 95 per cent thrust and other monitored systems indicate they are functioning properly. The arms release on receipt of a launch commit signal at the zero mark in the count. But the vehicle is prevented from accelerating too rapidly by controlled release mechanisms.

The Transporter

The six-million-pound transporters, the largest tracked vehicles known, move mobile launchers into the VAB and mobile launchers with assembled Apollo space vehicles to the launch pad. They also are used to transfer the mobile service structure to and from the launch pads. Two transporters are in use at Complex 39.

The Transporter is 131 feet long and 114 feet wide. The vehicle moves on four double-tracked crawlers, each 10 feet high and 40 feet long. Each shoe on the crawler tracks seven feet six inches in length and weighs about a ton.

Sixteen traction motors powered by four 1,000-kilowatt generators, which in turn are driven by two 2,750-horsepower diesel engines, provide the motive power for the transporter. Two 750-kw generators, driven by two 1,065-horsepower diesel engines, power the jacking, steering, lighting, ventilating and electronic systems.

Maximum speed of the transporter is about one-mile per-hour loaded and about two-miles per-hour unloaded. A 3½ mile trip to the pad with a mobile launcher, made at less than maximum speed, takes approximately seven hours.

The transporter has a leveling system designed to keep the top of the space vehicle vertical within plus-or-minus 10 minutes of arc -- about the dimensions of a basketball.

This system also provides leveling operations required to negotiate the five per cent ramp which leads to the launch pad, and keeps the load level when it is raised and lowered on pedestals both at the pad and within the VAB.

The overall height of the transporter is 20 feet from ground level to the top deck on which the mobile launcher is mated for transportation. The deck is flat and about the size of a baseball diamond (90 by 90 feet).

Two operator control cabs, one at each end of the chassis located diagonally opposite each other, provide totally enclosed stations from which all operating and control functions are coordinated.

The transporter moves on a roadway 131 feet wide, divided by a median strip. This is almost as broad as an eight-lane turnpike and is designed to accommodate a combined weight of about 18 million pounds.

The roadway is built in three layers with an average depth of seven feet. The roadway base layer is two-and-one-half feet of hydraulic fill compacted to 95 per cent density. The next layer consists of three feet of crushed rock packed to maximum density, followed by a layer of one foot of selected hydraulic fill. The bed is topped and sealed with an asphalt prime coat.

On top of the three layers is a cover of river rock, eight inches deep on the curves and six inches deep on the straightway. This layer reduces the friction during steering and helps distribute the load on the transporter bearings.

Mobile Service Structure

A 402-foot-tall, 9.8-million-pound tower is used to service the Apollo launch vehicle and spacecraft at the pad. The 40-story steel-trussed tower, called a mobile service structure, provides 360-degree platform access to the Saturn vehicle and the Apollo spacecraft.

The service structure has five platforms -- two self propelled and three fixed, but movable. Two elevators carry personnel and equipment between work platforms. The platforms can open and close around the 363-foot space vehicle.

After depositing the mobile launcher with its space vehicle on the pad, the transporter returns to a parking area about 7,000 feet from the pad. There it picks up the mobile service structure and moves it to the launch pad. At the pad, the huge tower is lowered and secured to four mount mechanisms.

The top three work platforms are located in fixed positions which serve the Apollo spacecraft. The two lower movable platforms serve the Saturn V.

The mobile service structure remains in position until about T-11 hours when it is removed from its mount and returned to the parking area.

Water Deluge System

A water deluge system will provide a million gallons of industrial water for cooling and fire prevention during launch of Apollo 9. Once the service arms are retracted at liftoff, a spray system will come on to cool these arms from the heat of the five Saturn F-1 engines during liftoff.

On the deck of the mobile launcher are 29 water nozzles. This deck deluge will start immediately after liftoff and will pour across the face of the launcher for 30 seconds at the rate of 50,000 gallons-per-minute. After 30 seconds, the flow will be reduced to 20,000 gallons-per-minute.

Positioned on both sides of the flame trench are a series of nozzles which will begin pouring water at 80,000 gallons-per-minute, 10 seconds before liftoff. This water will be directed over the flame deflector.

Other flush mounted nozzles, positioned around the pad, will wash away any fluid spill as a protection against fire hazards.

Water spray systems also are available along the egress route that the astronauts and closeout crews would follow in case an emergency evacuation was required.

Flame Trench and Deflector

The flame trench is 58 feet wide and approximately six feet above mean sea level at the base. The height of the trench and deflector is approximately 42 feet.

The flame deflector weighs about 1.3 million pounds and is stored outside the flame trench on rails. When it is moved beneath the launcher, it is raised hydraulically into position. The deflector is covered with a four-and-one-half-inch thickness of refractory concrete consisting of a volcanic ash aggregate and a calcium aluminate binder. The heat and blast of the engines are expected to wear about three-quarters of an inch from this refractory surface during the Apollo 9 launch.

Pad Areas

Both Pad A and Pad B of Launch Complex 39 are roughly octagonal in shape and cover about one fourth of a square mile of terrain.

The center of the pad is a hardstand constructed of heavily reinforced concrete. In addition to supporting the weight of the mobile launcher and the Saturn V vehicle, it also must support the 9.8-million-pound mobile service structure and 6-million-pound transporter, all at the same time. The top of the pad stands some 48 feet above sea level.

Saturn V propellants -- liquid oxygen, liquid hydrogen, and RP-1 -- are stored near the pad perimeter.

Stainless steel, vacuum-jacketed pipes carry the liquid oxygen (LOX) and liquid hydrogen from the storage tanks to the pad, up the mobile launcher, and finally into the launch vehicle propellant tanks.

LOX is supplied from a 900,000-gallon storage tank. A centrifugal pump with a discharge pressure of 320 pounds-per-square-inch pumps LOX to the vehicle at flow rates as high as 10,000-gallons-per-minute.

Liquid hydrogen, used in the second and third stages, is stored in an 850,000-gallon tank, and is sent through 1,500 feet of 10-inch vacuum-jacketed invar pipe. A vaporizing heat exchanger pressurizes the storage tank to 60 psi for a 10,000-gallons-per-minute flow rate.

The RP-1 fuel, a high grade of kerosene is stored in three tanks — each with a capacity of 86,000 gallons. It is pumped at a rate of 2,000 gallons-per-minute at 175 Psig.

The Complex 39 pneumatic system includes a converter compressor facility, a pad high-pressure gas storage battery, a high-pressure storage battery in the VAB, low and high pressure, cross-country supply lines, high-pressure hydrogen storage and conversion equipment, and pad distribution piping to pneumatic control panels. The various purging systems require 187,000 pounds of liquid nitrogen and 21,000 gallons of helium.

MISSION CONTROL CENTER

The Mission Control Center at the Manned Spacecraft Center, Houston, is the focal point for all Apollo flight control activities. The Center will receive tracking and telemetry data from the Manned Space Flight Network. These data will be processed through the Mission Control Center Real-Time Computer Complex and used to drive displays for the flight controllers and engineers in the Mission Operations Control Room and staff support rooms.

The Manned Space Flight Network tracking and data acquisition stations link the flight controllers at the Center to the spacecraft.

For Apollo 9, all stations will be remote sites without flight control teams. All uplink commands and voice communications will originate from Houston, and telemetry data will be sent back to Houston at high speed (2,400 bits per second), on two separate data lines. They can be either real time or playback information.

Signal flow for voice circuits between Houston and the remote sites is via commercial carrier, usually satellite, wherever possible using leased lines which are part of the NASA Communications Network.

Commands are sent from Houston to NASA's Goddard Space Flight Center, Greenbelt, Md., lines which link computers at the two points. The Goddard computers provide automatic switching facilities and speed buffering for the command data. Data are transferred from Goddard to remote sites on high speed (2,400 bits per second) lines. Command loads also can be sent by teletype from Houston to the remote sites at 100 words per minute. Again, Goddard computers provide storage and switching functions.

Telemetry data at the remote site are received by the RF receivers, processed by the Pulse Code Modulation ground stations, and transferred to the 642B remote-site telemetry computer for storage. Depending on the format selected by the telemetry controller at Houston, the 642B will output the desired format through a 2010 data transmission unit which provides parallel to serial conversion, and drives a 2,400 bit-per-second modem.

The data modem converts the digital serial data to phase-shifted keyed tones which are fed to the high speed data lines of the Communications Network.

Tracking data are output from the sites in a low speed (100 words) teletype format and a 240-bit block high speed (2,400 bits) format. Data rates are one sample-6 seconds for teletype and 10 samples (frames) per second for high speed data.

All high-speed data, whether tracking or telemetry, which originate at a remote site are sent to Goddard on high speed lines. Goddard reformats the data when necessary and sends them to Houston in 600-bit blocks at a 40,800 bits-per second rate. Of the 600-bit block, 480 bits are reserved for data, the other 120 bits for address, sync, inter-computer instructions, and polynomial error encoding.

All wideband 40,800 bits-per-second data originating at Houston are converted to high speed (2,400 bits-per-second) data at Goddard before being transferred to the designated remote site.

MANNED SPACE FLIGHT NETWORK

The Manned Space Flight Tracking Network for Apollo 9, consisting of 14 ground stations, four instrumented ships and six instrumented aircraft, is participating in its third manned flight. It is the global extension of the monitoring and control capability of the Mission Control Center in Houston. The network, developed by NASA through the Mercury and Gemini programs, now represents an investment of some $500 million and, during flight operations, has 4,000 persons on duty. In addition to NASA facilities. the network includes facilities of the Department of Defense and the Australian Department of Supply.

The network was developed by the Goddard Space Flight Center (GSFC) under the direction of NASA's Office of Tracking and Data Acquisition.

Basically, manned flight stations provide one or more of the following functions for flight control:

1. Telemetry
2. Tracking
3. Command and
4. Voice communications with the spacecraft

Apollo missions require the network to obtain information -- instantly recognize it, decode it, and arrange it for computer processing and display in the Mission Control Center.

Apollo generates much more information than either Projects Mercury or Gemini; therefore, very high speed data processing and display capability are needed. Apollo also requires network support at both Earth orbital and lunar distances. The Apollo Unified S-Band System (USB) provides this capability.

Network Configuration for Apollo 9
Unified S-Band Sites
NASA 30-Ft. Antenna Sites
Antigua (ANG)
Ascension Island (ACN)
Bermuda (BDA)
Canary Island (CYI)
Carnarvon (CRO), Australia
Grand Bahama Island (GBM)

Guam (OWM)
Guaymas (GYM), Mexico

Hawaii (HAW)
Merritt Island (MIL), Fla.

Texas (TEX), Corpus Christi

NASA 85-Ft. Antenna Sites
Honeysuckle Creek (HSK),
Australia (Prime)
Goldstone (GDS), Calif. (prime)

Madrid (MAD), Spain (Prime)
*Canberra (DSS-42-Apollo Wing)
(Backup)

*Goldstone (DSS-11-Apollo Wing
(Backup) MARS 210'-(Backup)

*Madrid (DSS-61-Apollo Wing)
(Backup)

Tananarive (TAN), Malagasy Republic (STADAN station in support role only.)

* Wings have been added to JPL Deep Space Network site operations buildings. These wings contain additional Unified S-Band equipment as backup to the Prime sites to allow for the use, if necessary, of the Deep Space 85-ft. antennas.

Network Testing

The MSFN began its Network Readiness Testing for Apollo 9 on Jan. 20 and continued through Feb. 5 when the network went on mission status. Through established computer techniques Goddard's Real-Time Computer Center (RTCC) conducted system-by-system, station-by-station tests of all tracking/data

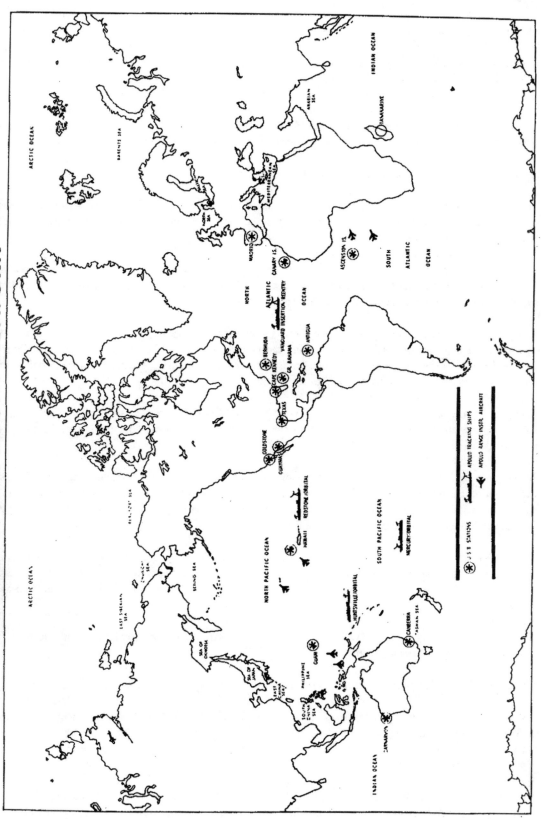

acquisition and communications systems until all support criteria were met and the MSFN pronounced ready to participate in the mission.

Spacecraft Communications

All Manned Space Flight Network stations are prepared to communicate with the Apollo 9 spacecraft in two different modes, S-Band and VHF.

When a station acquires the spacecraft, those sites having Unified S-Band and VHF air-to-ground capability will select the best quality and pass it to the Mission Control Center. All stations will monitor air-to-ground conversations for a possible crew request to switch the spacecraft communications.

NASA Communications Network - Goddard

This network consists of several systems of diversely routed communications channels leased on communications satellites, common carrier systems and high frequency radio facilities where necessary to provide the access links.

The system consists of both narrow and wide-band channels, and some TV channels. Included are a variety of telegraph, voice and data systems (digital and analog) with a wide range of digital data rates. Alternate routes or redundancy are provided for added reliability.

The primary switching center and intermediate switching and control points are established to provide centralized facility and technical control under direct NASA control. The primary switching center is at Goddard, and intermediate switching centers are located at Canberra, Australia; Madrid, Spain; London, England; Honolulu - Hawaii; Guam and Cape Kennedy, Fla.

Cape Kennedy is connected directly to the Mission Control Center by the communication network's Apollo Launch Data System (ALDS), a combination of data gathering and transmission systems designed to handle launch data exclusively.

After launch all network and tracking data are directed to the Mission Control Center through Goddard. A high-speed data line connects Cape Kennedy to Goddard, where the transmission rate is increased from there to Mission Control Center. Upon orbital insertion, tracking responsibility is transferred between the various stations as the spacecraft circles the Earth.

Two Intelsat communications satellites will be used for Apollo 9, one positioned over the Atlantic Ocean at about 60 degrees W. longitude in a near equatorial synchronous orbit varying about six degrees N. and S. in latitude. The Atlantic satellite will service the Ascension Island USB station, the Atlantic Ocean ship and the Canary Island site.

Only two of these three stations will be transmitting information back to Goddard at any one time, but all three stations can receive at all times.

The second Apollo Intelsat communications satellite is located about 170 degrees E. longitude over the mid-Pacific near the Equator at the international dateline. It will service the Carnarvon, Australian USB site and the Pacific Ocean ships. All these stations will be able to transmit simultaneously through the satellite to the Mission Control Center via Jamesburg, Calif., and the Goddard Space Flight Center.

Sites with "Dual" Capability

Certain stations of the Manned Space Flight Network can provide tracking, voice and data acquisition for two Apollo spacecraft simultaneously, provided they are within the beam width of the single Unified S-Band antenna. Two sets of frequencies separated by approximately five megahertz are used for this purpose. In addition to this primary mode of communications, the Unified S-Band system has the capability of receiving

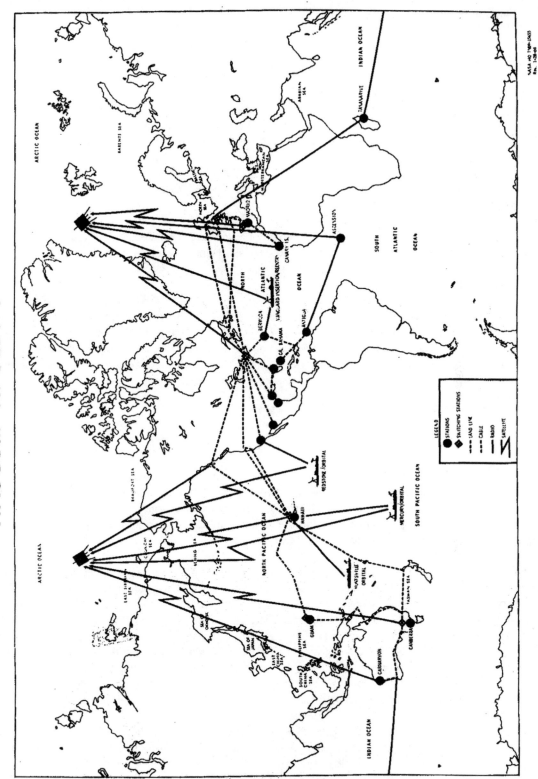

data on two other frequencies; primarily used for downlink data from the CSM.

For Apollo 9 this capability will be utilized when the LM and CSM have separated. Effective "dual" acquisition displacement distance between CSM and LM is one-hundred miles at a relative angle of forty-five degrees.

The "Dual" sites:
85 Ft. Antenna Systems
Honeysuckle Creek, Australia   Prime Site Wing Site (backup)
Madrid, Spain   Prime Site Wing Site (backup)
Goldstone, Calif.   Prime Site Wing Site (backup)
30 Ft. Antenna Systems
Carnarvon, Australia   Kauai, Hawaii
Ascension Island   Bermuda (Uplink only)
Merritt Island, Fla.   Antigua (Uplink only)
Guam Island, Pacific   USNS REDSTONE
                                                      USNS MERCURY
                                                      USNS VANGUARD

Spacecraft Television

Television transmissions will be received, recorded and converted to commercial (home) format for release
to the public by the Merritt Island, Fla., and Goldstone, Calif., Manned Space Flight Network stations.

Land Station Tracking

After the S-IVB and CSM/LM separate, dual-capability stations will be required to track both vehicles
simultaneously to provide HF A/G voice remoting from the CSM/LM and VHF TLM data from the S-IVB/IU
After LM power up, dual capability stations will be required to track the CSM via acquisition bus to provide
VHF voice, and to track the LM on VHF to provide VHF TLM or voice. Stations having only one VHF system
will track the CSM or LM in accordance with Houston requirements.

SHIPS AND AIRCRAFT NETWORK SUPPORT - APOLLO 9

The Apollo Instrumentation Ships (AIS)

The Apollo 9 mission will be supported by four Apollo Instrumentation Ships operating as integral stations
of the Manned Space Flight Network (MSFN) to provide coverage in areas beyond the range of land stations
The ships USNS Vanguard, Redstone, Mercury, and Huntsville will perform tracking, telemetry
communications, and computer functions for the launch and Earth orbit insertion phase, LM maneuver
phases, and re-entry at end of mission.

The Vanguard will be stationed due east of Bermuda (32 degrees N - 45 degrees W) during launch and will
remain at this position to cover specific orbital phases. The Vanguard also functions as part of the Atlantic
recovery fleet in the event of a launch phase contingency. The Redstone, Mercury, and Huntsville will all be
deployed in the Pacific Ocean area. The Redstone will be east of Hawaii (22 degrees N - 131 degrees W), the
Mercury southeast of American Samoa (22 degrees S 160 degrees W), and the Huntsville northwest of
American Samoa (7 degrees S -170 degrees E). The stations will provide communications, tracking, and
telemetry coverage in the areas where spacecraft and LM functions occur without adequate coverage from
land stations. Since the ships are a mobile instrumentation station, they can be repositioned prior to or during
the mission based upon flight plan needs subject to their speed limitations and weather conditions.

The Apollo ships were developed jointly by NASA and the Department of Defense. The DOD operates the
ships in support of Apollo and other NASA/DOD missions on a noninterference basis with Apollo support
requirements. The overall management of the Apollo ship operation is the responsibility of the Air Force
Western Test Range (AFWTR). The Military Sea Transport Service MSTS) provides the maritime crews; the
Federal Electric Co. (FEC) (under contract to AFWTR) provides the technical instrumentation crews.
Goddard Space Flight Center (GSFC) is responsible for the configuration control and network interface of
the ships in support of Apollo missions. The technical crews operate in accordance with joint NASA/DOD
technical standards and specifications which are compatible with the manned space flight network
procedures.

The Apollo Range Instrumentation Aircraft (ARIA)

The ARIA will support the Apollo 9 mission by filling gaps in critical coverage beyond the range of land and ship stations or at points where land or ship coverage is either impossible or impractical. During this mission six ARIA will be used in the Pacific and Atlantic areas to receive and record telemetry from the S-IVB stage, and later on in the mission from the CSM/LM. In addition, the ARIA will provide a two-way voice relay between the Mission Control Center and the Apollo 9 crew during critical maneuver periods.

On launch day, four ARIA will be used in the Pacific area during revolutions 2, 3 and 4 to cover LM/CSM/S-IVB functions. These aircraft will use staging bases in Australia, Guam, and Hawaii. On day two, revolution 27, one ARIA will deploy to a point about 400 miles ENE of Darwin, Australia, and a second ARIA will deploy to a point some 900 miles SSE of Tahiti to provide critical telemetry and voice relay coverage during activation of LM electrical propulsion and environmental control systems. Staging bases for this coverage will be Darwin and Pago Pago. On day four, two ARIA will be deployed in the Atlantic area northeast and northwest of Ascension Island to provide active voice relay at predetermined points on revolutions 61, 62, and 63 during CSM/LM maneuvers. At the end of the mission, at least one ARIA will be positioned south of Bermuda to provide telemetry and voice relay from the CSM from the 400K foot level to splashdown. ARIA operating in the Atlantic area will use bases at Ascension Island Puerto Rico, and Patrick AFB, Florida.

The total ARIA fleet consists of eight EC-135N (Boeing 707) jet aircraft equipped specifically to meet mission needs. Seven foot diameter parabolic (dish) antennas have been installed in the nose section of each aircraft giving them a large, bulbous look. The ARIA airframes were selected from the USAF transport fleet, modified through joint DOD/NASA contract, and are operated and maintained by the Air Force Eastern Test Range (AFETR). During their support of Apollo missions they become an integral part of the MSFN and, as such, operate in accordance with the joint NASA/DOD procedures and policies which govern the MSFN.

APOLLO 9 CREW

Crew Training

The crewmen of Apollo 9 will have spent more than seven hours of formal training for each hour of the mission's 10-day duration. Almost 1,800 hours of training were in the Apollo 9 crew training syllabus over and above the normal preparations for the mission -- technical briefings and reviews, pilot meetings and study.

The Apollo 9 crewmen also took part in spacecraft manufacturing checkouts at the North American Rockwell plant in Downey, Calif., at Grumman Aircraft Engineering Corp., Bethpage, N.Y., and in pre-launch testing at NASA Kennedy Space Center. Taking part in factory and launch area testing has provided the crew with thorough operational knowledge of the complex vehicle,

Highlights of specialized Apollo 9 crew training topics are:

* Detailed series of briefings on spacecraft systems, operation and modifications.

* Saturn launch vehicle briefings on countdown, range safety, flight dynamics, failure modes and abort conditions. The launch vehicle briefings were updated periodically.

* Apollo Guidance and Navigation system briefings at the Massachusetts Institute of Technology Instrumentation Laboratory.

* Briefings and continuous training on mission photographic objectives and use of camera equipment.

* Extensive pilot participation in reviews of all flight procedures for normal as well as emergency situations.

* Stowage reviews and practice in training sessions in the spacecraft, mockups, and Command Module simulators allowed the crewmen to evaluate spacecraft stowage of crew-associated equipment.

* Zero-g aircraft flights using command module and lunar module mockups for EVA and pressure suit doffing/donning practice and training.

* Underwater zero-g training in the MSC Water Immersion Facility using spacecraft mockups to further familiarize crew with all aspects of CSM-LM docking tunnel intravehicular transfer and EVA in pressurized suits.

* More than 300 hours of training per man in command module and lunar module mission simulators at MSC and KSC, including closed-loop simulations with flight controllers in the Mission Control Center. Other Apollo simulators at various locations were used extensively for specialized crew training.

* Water egress training conducted in indoor tanks as well as in the Gulf of Mexico included uprighting from the Stable II position (apex down) to the Stable I position (apex up), egress onto rafts and helicopter pickup.

* Launch pad egress training from mockups and from the actual spacecraft on the launch pad for possible emergencies such as fire, contaminants and power failures.

* The training covered use of Apollo spacecraft fire suppression equipment in the cockpit.

* Planetarium reviews at Morehead Planetariums Chapel Hill, N. C., and at Griffith Planetarium, Los Angeles, of the celestial sphere with special emphasis on the 37 navigational stars used by the Command Module Computer.

Crew Life Support Equipment

Apollo 9 crewmen will wear two versions of the Apollo spacesuit: an intravehicular pressure garment assembly worn by the command module pilot and the extravehicular pressure garment assembly worn by the commander and the lunar module pilot. Both versions are basically identical except that the extravehicular version has an integral thermal meteroid garment over the basic suit.

From the skin out, the basic pressure garment consists of a nomex comfort layer, a neoprene-coated nylon pressure bladder and a nylon restraint layer. The outer layers of the intravehicular suit are, from the inside out, nomex and two layers of Teflon-coated Beta cloth. The extravehicular integral thermal meteoroid cover consists of a liner of two layers of neoprene-coated nylon, seven layers of Beta/Kapton spacer laminate, and an outer layer of Teflon-coated Beta fabric.

The extravehicular suit, together with a liquid cooling garment, portable life support system (PLSS), oxygen purge system, extravehicular visor assembly and other components make up the extravehicular mobility unit (EMU). The EMU provides an extravehicular crewman with life support for a four hour mission outside the lunar module without replenishing expendables. EMU total weight is 183 pounds. The intravehicular suit weighs 35.6 pounds.

Liquid Cooling Garment -- A knitted nylon-spandex garment with a network of plastic tubing through which cooling water from the PLSS is circulated. It is worn next to the skin and replaces the constant wear garment during EVA only.

Portable life support system -- A backpack supplying oxygen at 3.9 psi and cooling water to the liquid cooling garment. Return oxygen is cleansed of solid and gas contaminants by a lithium hydroxide canister. The PLSS includes communications and telemetry equipment, displays and controls and a main power supply. The PLSS is covered by a thermal insulation jacket.

Oxygen Purge System -- Mounted atop the PLSS, the oxygen purge system provides a contingency 30-minute supply of gaseous oxygen in two two-pound bottles pressurized to 5,880 psia. The system may also be worn separately on the front of the pressure garment assembly torso. It serves as a mount for the VHF antenna for the PLSS.

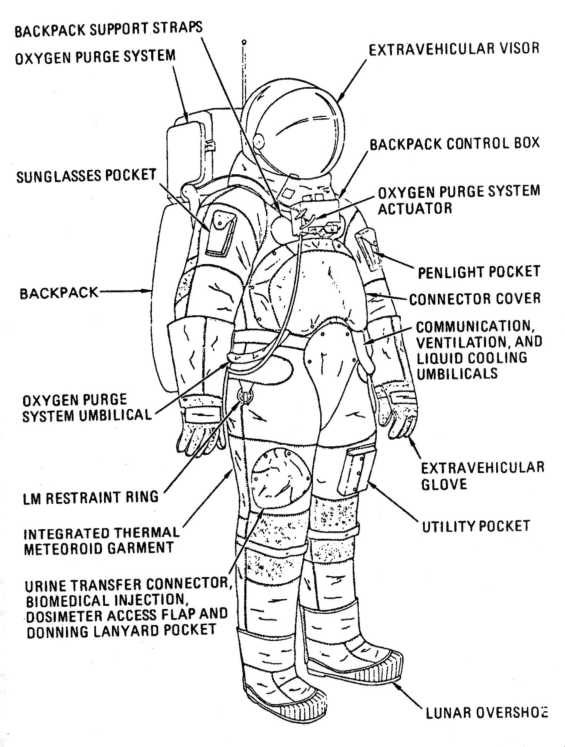

BACKPACK SUPPORT STRAPS

OXYGEN PURGE SYSTEM

EXTRAVEHICULAR VISOR

BACKPACK CONTROL BOX

SUNGLASSES POCKET

OXYGEN PURGE SYSTEM ACTUATOR

BACKPACK

PENLIGHT POCKET

CONNECTOR COVER

COMMUNICATION, VENTILATION, AND LIQUID COOLING UMBILICALS

OXYGEN PURGE SYSTEM UMBILICAL

LM RESTRAINT RING

EXTRAVEHICULAR GLOVE

UTILITY POCKET

INTEGRATED THERMAL METEOROID GARMENT

URINE TRANSFER CONNECTOR, BIOMEDICAL INJECTION, DOSIMETER ACCESS FLAP AND DONNING LANYARD POCKET

LUNAR OVERSHOE

Extravehicular visor assembly -- A polycarbonate shell and two visors with thermal control and optical coatings on them. The EVA visor is attached over the pressure helmet to provide impact, micrometeoroid, thermal and light protection to the EVA crewman.

Extravehicular Gloves -- Built of an outer shell of Chromel-R fabric and thermal insulation to provide protection when handling extremely hot and cold objects. The finger tips are made of silicone rubber to provide the crewman more sensitivity.

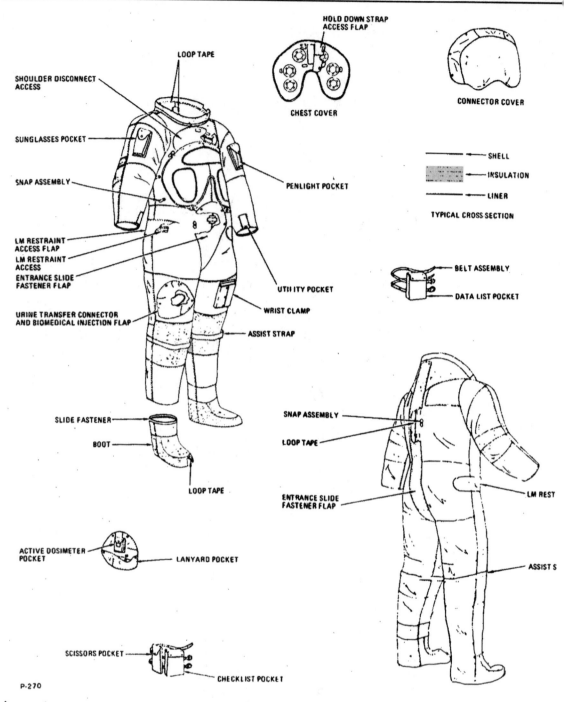

HOLD DOWN STRAP
ACCESS FLAP

LOOP TAPE

SHOULDER DISCONNECT
ACCESS

CHEST COVER

CONNECTOR COVER

SUNGLASSES POCKET

SHELL

INSULATION

LINER

SNAP ASSEMBLY

PENLIGHT POCKET

TYPICAL CROSS SECTION

LM RESTRAINT
ACCESS FLAP
LM RESTRAINT
ACCESS
ENTRANCE SLIDE
FASTENER FLAP

BELT ASSEMBLY

DATA LIST POCKET

UTILITY POCKET

URINE TRANSFER CONNECTOR
AND BIOMEDICAL INJECTION FLAP

WRIST CLAMP

ASSIST STRAP

SLIDE FASTENER

BOOT

SNAP ASSEMBLY

LOOP TAPE

LOOP TAPE

LM REST

ENTRANCE SLIDE
FASTENER FLAP

ACTIVE DOSIMETER
POCKET

LANYARD POCKET

ASSIST S

SCISSORS POCKET

CHECKLIST POCKET

P-270

A one-piece constant wear garment, similar to "long johns" is worn as an undergarment for the spacesuit i
intravehicular operations and for the in-flight coveralls. The garment is porous-knit cotton with a
waist-to-neck zipper for donning. Biomedical harness attach points are provided.

During periods out of the spacesuits, crewmen will wear two-piece Teflon fabric in-flight coveralls for warmth
and for pocket stowage of personal items.

Communications carriers ("Snoopy hats") with redundant microphones and earphones are worn with the
pressure helmet; a lightweight headset is worn with the in-flight coveralls.

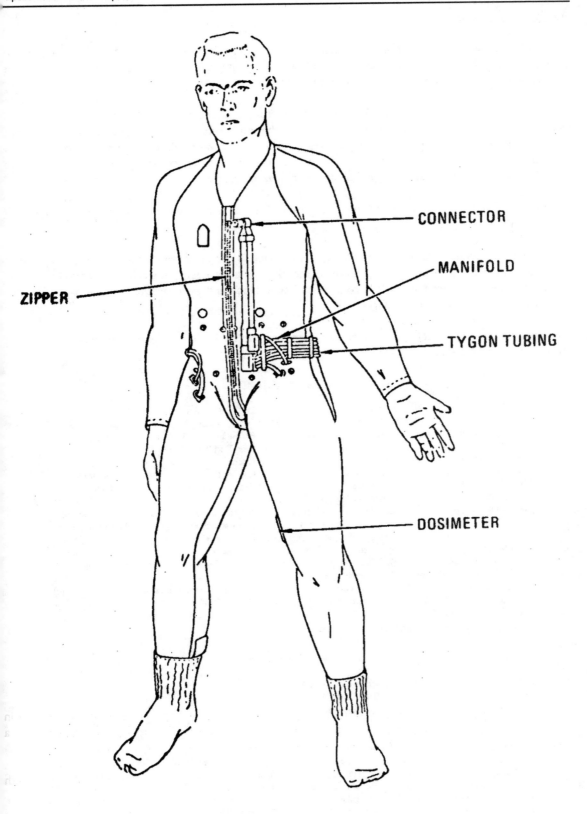

CONNECTOR

MANIFOLD

ZIPPER

TYGON TUBING

DOSIMETER

Apollo 9 Crew Meals

The Apollo 9 crew has a wide range of food items from which to select their daily mission space menu. More than 60 items comprise the food selection list of freeze-dried bite-size and re-hydratable foods. The average daily value of three meals will be 2,500 calories per man.

Water for drinking and rehydrating food is obtained from three sources in the command module -- a dispenser for drinking water and two water spigots at the food preparation station, one supplying water at about 155 degrees F, the other at about 55 degrees F. The potable water dispenser emits half-ounce spurts with each squeeze and the food preparation spigots dispense water in I-ounce increments. Command module potable water is supplied from service module fuel cell by-product water.

A similar hand water dispenser aboard the lunar module is used for cold-water re-hydration of food packets stowed in the LM.

After water has been injected into a food bag, it is kneaded for about three minutes. The bag neck is then cut off and the food squeezed into the crewman's mouth. After a meal, germicide pills attached to the outside of the food bags are placed in the bags to prevent fermentation and gas formation and the bags are rolled and stowed in waste disposal compartments.

The day-by-day, meal-by-meal Apollo 9 menu for each crewman for both the command module and the lunar module is listed on the following pages.

APOLLO 9 - McDivitt

MEAL	DAY 1*, 5, 9	DAY 2, 6, 10	DAY 3, 7, 11	DAY 4, 8
A	Peaches	Canadian Bacon & Applesauce	Fruit Cocktail	Sausage Patties
	Bacon Squares (8)	Sugar Coated Corn Flakes	Bacon Squares (8)	Peaches
	Cinn Tstd Bread Cubes (8)	Brownies (8)	Cinn Tstd Bread Cubes (8)	Bacon Squares (8)
	Grapefruit Drink	Grapefruit Drink	Cocoa	Cocoa
	Orange Drink	Grape Drink	Orange Drink	Grape Drink
B	Salmon Salad	Tuna Salad	Cream of Chicken Soup	Pea Soup
	Chicken & Gravy	Chicken & Vegetables	Beef Pot Roast	Chicken & Gravy
	Toasted Bread Cubes (6)	Cinn Tstd Bread Cubes (8)	Toasted Bread Cubes (8)	Cheese Sandwiches (6)
	Sugar Cookie Cubes (6)	Pineapple Fruitcake (4)	Butterscotch Pudding	Bacon Squares (6)
	Cocoa	Pineapple-Grapefruit Drink	Grapefruit Drink	Grapefruit Drink
C	Beef & Gravy	Spaghetti & Meat Sauce	Beef Hash	Shrimp Cocktail
	Beef Sandwiches (4)	Beef Bites (6)	Chicken Salad	Beef & Vegetables
	Cheese-Cracker Cubes (8)	Bacon Squares (6)	Turkey Bites (6)	Cinn Tstd Bread Cubes (8)
	Chocolate Pudding	Banana Pudding	Graham Cracker Cubes (6)	Date Fruitcake (4)
	Orange-Grapefruit Drink	Grapefruit Drink	Orange Drink	Orange-Grapefruit Drink

*Day I consists of Meals B and C only

Each crewmember will be provided with a total of 32 meals

APOLLO 9 - Scott

MEAL	DAY 1*, 5, 9	DAY 2, 6, 10	DAY 3, 7, 11	DAY 4, 8
A	Peaches	Canadian Bacon & Applesauce	Fruit Cocktail	Sausage Patties
	Bacon Squares (8)	Sugar coated Corn Flakes	Bacon Squares (8)	Peaches
	Cinn Tstd Bread Cubes (8)	Brownies (8)	Cinn Tstd Bread Cubes (8)	Bacon Squares (8)
	Grapefruit Drink	Grapefruit Drink	Cocoa	Cocoa
	Orange Drink	Grape Drink	Orange Drink	Grape Drink
B	Salmon Salad	Tuna Salad	Cream of Chicken Soup	Pea Soup
	Chicken & Gravy	Chicken & Vegetables	Beef Pot Roast	Chicken & Gravy
	Toasted Bread Cubes (6)	Cinn.Tstd Bread Cubes (8)	Toasted Bread Cubes (8)	Cheese Sandwiches (6)
	Sugar Cookie Cubes (6)	Pineapple Fruitcake (4)	Butterscotch Pudding	Bacon Squares (6)
	Cocoa	Pineapple-Grapefruit Drink	Grapefruit Drink	Grapefruit Drink
C	Beef & Gravy	Spaghetti & Meat Sauce	Beef Hash	Shrimp Cocktail
	Beef Sandwiches (4)	Beef Bites (6)	Chicken Salad	Beef & Vegetables
	Cheese-Cracker Cubes (8)	Bacon Squares (6)	Turkey Bites (6)	Cinn Tstd Bread Cubes (8)
	Chocolate Pudding	Banana Pudding	Graham Cracker Cubes (6)	Date Fruitcake (4)
	Orange-Grapefruit Drink	Grapefruit Drink	Orange Drink	Orange-Grapefruit Drink

*Day I consists of Meals B and C only
Each crewmember will be provided with a total of 32 meals

APOLLO 9 - Schweickart

MEAL	DAY 1*, 5, 9	DAY 2, 6, 10	DAY 3, 7, 3.1	DAY 4, 8
A	Peaches	Canadian Bacon & Applesauce	Fruit Cocktail	Sausage Patties
	Bacon Squares (8)	Sugar Coated Corn Flakes	Bacon Squares (8)	Peaches
	Cinn Tstd Bread Cubes (8)	Brownies (8)	Cinn Tstd Bread Cubes (8)	Bacon Squares (8)
	Grapefruit Drink	Grapefruit Drink	Cocoa	Cocoa
	Orange Drink	Grape Drink	Orange Drink	Grape Drink
B	Salmon Salad	Tuna Salad	Cream of Chicken Soup	Pea Soup
	Chicken & Gravy	Chicken & Vegetables	Beef Pot Roast	Chicken & Gravy
	Toasted Bread Cubes (6)	Cinn Tstd Bread Cubes (8)	Toasted Bread Cubes (8)	Cheese Sandwiches (6)
	Sugar Cookie Cubes (6)	Pineapple Fruitcake (4)	Butterscotch Pudding	Bacon Squares (6)
	Cocoa	Pineapple-Grapefruit Drink	Grapefruit Drink	Grapefruit Drink
C	Beef & Gravy	Spaghetti & Meet Sauce	Beef Hash	Spaghetti & Heat Sauce
	Beef Sandwiches (4)	Beef Bites (6)	Chicken Salad	Beef & Vegetables
	Cheese-Cracker Cubes (8)	Bacon Squares (6)	Turkey Bites (6)	Cinn Tstd Bread Cubes (8)
	Chocolate Pudding	Banana Pudding	Graham Cracker Cubes (6)	Date' Fruitcake (4)
	Orange-Grapefruit Drink	Grapefruit Drink	Orange Drink	Orange-Grapefruit Drink

*Day I consists of Meals B and C only
Each crewmember will be provided with a total of 32 meals

APOLLO 9 LM MENU

Day I

Meal A

Chicken and Gravy
Butterscotch Pudding
Sugar Cookie Cubes (6)
Orange-Pineapple Drink
Grape Drink

Meal B

Chicken Salad
Beef Sandwiches (6)
Date Fruitcake (4)
Chocolate Pudding
Orange Drink

Meal C

Beef Hash
Bacon Squares (8)
Strawberry Cereal Cubes (6)
Pineapple Grapefruit Drink
Grape Drink

2 man-days are required
Red and Blue Velcro
2 meals per overwrap

Personal Hygiene

Crew personal hygiene equipment aboard Apollo 9 includes body cleanliness items, the waste management system and two medical kits.

Packaged with the food are a toothbrush and a 2-ounce tube of toothpaste for each crewman. Each man-meal package contains a 3.5 by 4-inch wet-wipe cleansing towel. Additionally, three packages of 12 by 12-inch dry towels are stowed beneath the command module pilot's couch. Each package contains seven towels. Also stowed under the command module pilot's couch are seven tissue dispensers containing 53 three-ply tissues each.

Solid body wastes are collected in Gemini-type plastic defecation bags which contain a germicide to prevent bacteria and gas formation. The bags are sealed after use and stowed in empty food containers for post-flight analysis.

Urine collection devices are provided for use either wearing the pressure suit or while in the in-flight coveralls. The urine is dumped overboard through the spacecraft urine dump valve in the CM and stored in the LM.

The two medical accessory kits, 6 by 4.5 by 4 inches, are stowed on the spacecraft back wall at the feet of the command module pilot.

The medical kits contain three motion sickness injectors, three pain suppression injectors, one 2-ounce bottle first aid ointment, two 1-ounce bottle eye drops, three nasal sprays, two compress bandages, 12 adhesive bandages, one oral thermometer and two spare crew biomedical harnesses. Pills in the medical kits are 60 antibiotic, 12 nausea, 12 stimulant, 18 pain killer, 60 decongestant, 24 diarrhea, 72 aspirin and 21 sleeping. Additionally, a small medical kit containing two stimulant and four pain killer pills is stowed in the lunar module food compartment.

Survival Gear

The survival kit is stowed in two rucksacks in the right-hand forward equipment bay above the lunar module pilot.

Contents of rucksack No. 1 are: two combination survival lights, one desalter kit, three pairs of sunglasses, one radio beacon, one spare radio beacon battery and spacecraft connector cable, one knife in sheath, three water containers and two containers of Sun lotion.

Rucksack No. 2: one three-man life raft with CO2 inflater, one sea anchor, two sea dye markers, three sunbonnets, one mooring lanyard, three manlines and two attach brackets.

The survival kit is designed to provide a 48-hour post-landing (water or land) survival capability for three crewmen between 40 degrees North and South latitudes.

Biomedical Inflight Monitoring

The Apollo 9 crew biomedical telemetry data received by the Manned Space Flight Network will be relayed for instantaneous display at Mission Control Center where heart rate and breathing rate data will be displayed on the flight surgeon's console. Heart rate and respiration rate average, range and deviation are computed and displayed on digital TV screens.

In addition, the instantaneous heart rate, real time and delayed EKG and respiration are recorded on strip charts for each man.

Biomedical telemetry will be simultaneous from all crewmen while in the CSM, but selectable by a manual onboard switch in the LM. During EVA the flight surgeons will be able to monitor the EVA LM pilot as well as the commander in the LM and the command module pilot in the CSM.

Biomedical data observed by the MOCR flight surgeon and his team in the Life Support Systems Staff Support Room will be correlated with spacecraft and spacesuit environmental data displays.

Blood pressures are no longer telemetered as they were in the Mercury and Gemini programs. Oral temperatures however, can be measured onboard for diagnostic purposes and voiced down by the crew in case of in-flight illness.

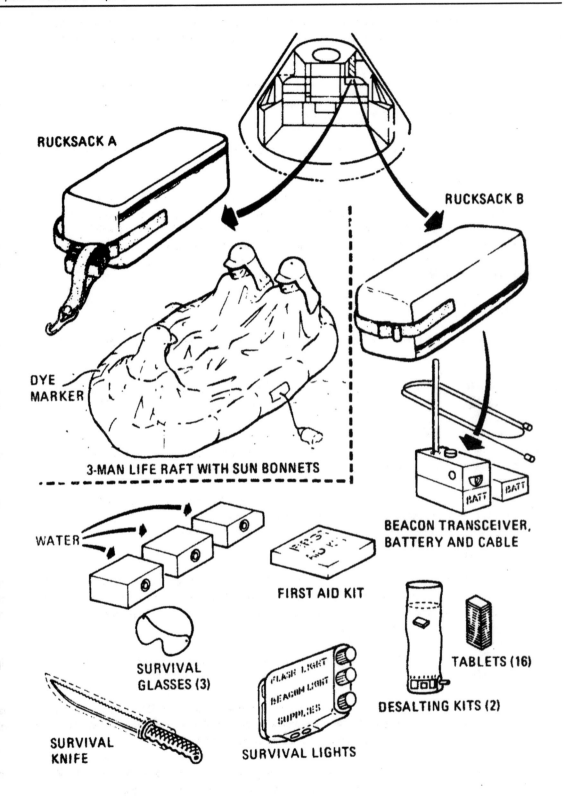

RUCKSACK A

RUCKSACK B

DYE MARKER

3-MAN LIFE RAFT WITH SUN BONNETS

BEACON TRANSCEIVER, BATTERY AND CABLE

WATER

FIRST AID KIT

SURVIVAL GLASSES (3)

TABLETS (16)

DESALTING KITS (2)

SURVIVAL KNIFE

SURVIVAL LIGHTS

FLASH LIGHT
BEACON LIGHT
SUPPLIES

Crew Launch-Day Timeline

Following is a timetable of Apollo 9 crew activities on launch day. (All times are shown in hours and minutes before liftoff.)

T-9:00   Backup crew alerted

T-8:30   Backup crew to LC-39A for spacecraft pre-launch checkouts

T-5:00   Flight crew alerted

T-4:45   Medical examinations

T-4:15   Breakfast

T-3:45   Don pressure suits

T-3:30   Leave Manned Spacecraft Operations Building for LC-39A via crew transfer van

T-3:14   Arrive at LC-39A

T-3:10   Enter elevator to spacecraft level

T-2:40   Begin spacecraft ingress

Rest-Work Cycles

All three Apollo 9 crewmen will sleep simultaneously during rest periods. The commander and the command module pilot will sleep in the couches and the lunar module pilot will sleep in the lightweight sleeping bag beneath the couches. Since each day's mission activity is of variable length, rest periods will not come at regular intervals.

During rest periods, both the commander and the command module pilot will wear their communications headsets and remain on alert duty, but with receiver volume turned down.

When possible, all three crewmen will eat together in 1-hour eat periods during which other activities will be held to a minimum.

Crew Biographies

NAME: James A. McDivitt (Colonel, USAF) Spacecraft Commander

BIRTHPLACE AND DATE: Born June 10, 1929, in Chicago, Ill. His parents, Mr. and Mrs. James McDivitt, reside in Jackson, Mich.

PHYSICAL DESCRIPTION: Brown hair; blue eyes; height: 5 feet 11 inches; weight: 155 pounds.

EDUCATION: Graduated from Kalamazoo Central High School, Kalamazoo, Mich.; received a Bachelor of Science degree in Aeronautical Engineering from the University of Michigan (graduated first in class) in 1959 and an Honorary Doctorate in Astronautical Science from the University of Michigan in 1965.

MARITAL STATUS: Married to the former Patricia A. Haas of Cleveland, OH. Her parents, Mr. and Mrs. William Haas, reside in Cleveland.

CHILDREN: Michael A., Apr. 14, 1957; Ann L., July 21, 1958; Patrick W., Aug. 30, 1960; Kathleen M., June 16, 1966.

OTHER ACTIVITIES: His hobbies include handball, hunting, golf, swimming, water skiing and boating.

ORGANIZATIONS: Member of the Society of Experimental Test Pilots, the American Institute of Aeronautics and Astronautics, Tau Beta Pi and Phi Kappa Phi.

SPECIAL HONORS: Awarded the NASA Exceptional Service Medal and the Air Force Astronaut Wings; four Distinguished Flying Crosses; five Air Medals; the Chong Moo Medal from South Korea; the USAF Air Force Systems Command Aerospace Primus Award; the Arnold Air Society JFK Trophy; the Sword of Loyola; and the Michigan Wolverine Frontiersman Award.

EXPERIENCE: McDivitt joined the Air Force in 1951 and holds the rank of Colonel. He flew 145 combat missions during the Korean War in F-80s and F-86s.

He is a graduate of the USAF Experimental Test Pilot School and the USAF Aerospace Research Pilot course and served as an experimental test pilot at Edwards Air Force Base, Calif.

He has logged 3,922 hours of flying time -- 3,156 hours in jet aircraft.

Colonel McDivitt was selected as an astronaut by NASA in September 1962.

He was command pilot for Gemini 4, a 66-orbit 4-day mission that began on June 3 and ended on June 7, 1965. Highlights of the mission included a controlled extravehicular activity period performed by pilot Ed White, cabin depressurization and opening of spacecraft cabin doors, and the completion of 12 scientific and medical experiments.

NAME: David R. Scott (Colonel USAF) Command Module Pilot

BIRTHPLACE AND DATE: Born June 6, 1932, in San Antonio, Tex. His parents, Brigadier Gen. (USAF Ret.) and Mrs. Tom W. Scott, reside in La Jolla, Calif.

PHYSICAL DESCRIPTION: Blond hair; blue eyes; height: 6 feet; weight: 175 pounds.

EDUCATION: Graduated from Western High School, Washington, D.C.; received a Bachelor of Science degree from the U. S. Military Academy and the degree of Master of Science in Aeronautics and Astronautics from the Massachusetts Institute of Technology.

MARITAL STATUS: Married to the former Ann Lurton Ott of San Antonio, Tex. Her parents are Brigadier Gen. (USAF Ret.) and Mrs. Isaac W. Ott of San Antonio.

CHILDREN: Tracy L., Mar. 25, 1961; Douglas W., Oct. 8, 1963.

OTHER ACTIVITIES: His hobbies are swimming, handball, skiing and photography.

ORGANIZATIONS: Associate Fellow of the American Institute of Aeronautics and Astronautics; member of the Society of Experimental Test Pilots; Tau Beta Pi; Sigma Xi; and Sigma Gamma Tau.

SPECIAL HONORS: Awarded the NASA Exceptional Service Medal, the Air Force Astronaut Wings, and the Distinguished Flying Cross; and recipient of the AIAA Astronautics Award.

EXPERIENCE: Scott graduated fifth in a class of 633 at West Point and subsequently chose an Air Force career. He completed pilot training at Webb Air Force Base, Tex., in 1955, and then reported for gunnery training at Laughlin Air Force Base, Tex., and Luke Air Force Base, Ariz.

He was assigned to the 32nd Tactical Fighter Squadron at Soesterberg Air Force Base (RNAF), Netherlands, from Apr. 1956 to July 1960. Upon completing this tour of duty, he returned to the United States for study at the Massachusetts Institute of Technology where he completed work on his Master's degree. His thesis at MIT concerned interplanetary navigation.

After completing his studies at MIT in June 1962, he attended the Air Force Experimental Test Pilot School and then the Aerospace Research Pilot School.

He has logged more than 3,800 hours flying time - 3,600 hours in jet aircraft.

Col. Scott was one of the third group of astronauts named by NASA in Oct. 1963.

On Mar. 16, 1966, he and command pilot Neil Armstrong were launched into space on the Gemini 8 mission -- a flight originally scheduled to last three days but terminated early due to a malfunctioning OAMS thruster. The crew performed the first successful docking of two vehicles in space and demonstrated great piloting skill in overcoming the thruster problem and bringing the spacecraft to a safe landing.

NAME: Russell L. Schweickart (Mr.) Lunar Module Pilot

BIRTHPLACE AND DATE: Born Oct. 25, 1935, in Neptune, N. J. His parents, Mr. and Mrs. George Schweickart. reside in Sea Girt, N.J.

PHYSICAL DESCRIPTION: Red hair; blue eyes; height: 6 feet; weight 161 pounds.

EDUCATION: Graduated from Manasquan High School, N.J.; received a Bachelor of Science degree in Aeronautical Engineering and a Master of Science degree in Aeronautics and Astronautics from Massachusetts Institute of Technology.

MARITAL STATUS: Married to the former Clare G. Whitfield of Atlanta, Ga. Her parents are the Randolph Whitfields of Atlanta.

CHILDREN: Vicki, Sept. 12, 1959; Randolph and Russell, Sept. 8, 1960; Elin, Oct. 19, 1961; Diana, July 26, 1964.

OTHER ACTIVITIES: His hobbies are amateur astronomy, photography and electronics.

ORGANIZATIONS: Member of the Sigma Xi.

EXPERIENCE: Schweickart served as a pilot in the United States Air Force and Air National Guard from 1956 to 1963.

He was a research scientist at the Experimental Astronomy Laboratory at MIT and his work there included research in upper atmospheric physics, star tracking and stabilization of stellar images. His thesis for a Master's degree at MIT concerned stratospheric radiance.

Of the 2,400 hours flight time he has logged, 2,100 hours are in jet aircraft.

Schweickart was one of the third group of astronauts named by NASA in Oct. 1963.

APOLLO PROGRAM MANAGEMENT/CONTRACTORS

Direction of the Apollo Program, the United States' effort to land men on the Moon and return them safely to Earth before 1970, is the responsibility of the Office of Manned Space Flight (OMSF), National Aeronautics and Space Administration, Washington, D.C. Dr. George E. Mueller is Associate Administrator for Manned Space Flight.

NASA Manned Spacecraft Center (MSC), Houston, is responsible for development of the Apollo spacecraft, flight crew training and flight control. Dr. Robert R. Gilruth is Center Director.

NASA Marshall Space Flight Center (MSFC), Huntsville, Ala., is responsible for development of the Saturn launch vehicles. Dr. Wernher von Braun is Center Director.

NASA John F. Kennedy Space Center (KSC), Fla., is responsible for Apollo/Saturn launch operations. Dr. Kurt H. Debus is Center Director.

NASA Goddard Space Flight Center (GSFC), Greenbelt, Md., manages the Manned Space Flight Network under the direction of the NASA Office of Tracking and Data Acquisition (OTDA). Gerald M. Truszynski is Associate Administrator for Tracking and Data Acquisition. Dr. John F. Clark is Director of GSFC.

Apollo/Saturn Officials

NASA Headquarters
Lt. Gen. Sam C. Phillips, (USAF)	Apollo Program Director, OMSF
George H. Rage	Apollo Program Deputy Directors Mission Director OMSF
Chester M. Lee	Assistant Mission Director OMSF
Col. Thomas H. McMullen (USAF)	Assistant Mission Director OMSF
Maj. Gen. James W. Humphreys, Jr.	Director of Space Medicine, OMSF
Worman Pozinsky	Director Network Support Implementation Div., OTDA
George M. Low	Manager, Apollo Spacecraft Program
Kenneth S. Kleinknecht	Manager, Command and Service Modules
Brig. Gen. C. H. Bolender (USAF)	Manager, Lunar Module
Donald K. Slayton	Director of Flight Crew Operations
Christopher C. Kraft, Jr.	Director of Flight Operations
Eugene F. Kranz	Flight Director
Gerald Griffin	Flight Director
M. P. Frank	Flight Director
Charles A. Berry	Director of Medical Research and operations

Marshall Space Flight Center
Maj. Gen. Edmund F. O'Connor	Director of Industrial Operations
Dr. F. A. Speer	Director of Mission Operations
Lee B. James	Manager, Saturn V Program Office
William D. Brown	Manager, Engine Program Office

Kennedy Space Center
Miles Ross	Deputy Director, Center Operations
Rear Adm. Roderick O. Middleton (USN)	Manager, Apollo Program Office
Rocco A. Petrone	Directors Launch operations
Walter J. Kapryan	Deputy Director, Launch Operations
Dr. Hans F. Gruene	Director, Launch Vehicle Operations
John J. Williams	Director, Spacecraft Operations
Paul C. Donnelly	Launch Operations Manager

Goddard Space Flight Center
Ozro M. Covington	Assistant Director for Manned Space Flight Tracking
Henry F. Thompson	Deputy Assistant Director for Manned Space Flight Support
H. William Wood	Chief, Manned Flight Operations Div.
Tecwyn Roberts	Chief, Manned Flight Engineering Div.

Department of Defense
Maj. Gen. Vincent G. Huston, (USAF)	DOD Manager of Manned Space Flight Support Operations
Maj. Gen. David M, Jones, (USAF)	Deputy DOD Manager of Manned Space Flight Support Operations, Commander of USAF Eastern Test Range
Rear Adm. P. S. McManus, (USN)	Commander of Combined Task Force 140, Atlantic Recovery Area
Rear Adm. F. E. Bakutis, (USN)	Commander of Combined Task Force 130, Pacific Recovery Area
Col. Royce G. Olson, (USAF)	Director of DOD Manned Space Flight Office
Brig. Gen. Allison C. Brooks, (USAF)	Commander Aerospace Rescue and Recovery Service

Major Apollo/Saturn V Contractors

Contractor	Item
Bellcomm Washington, D.C.	Apollo Systems Engineering
The Boeing Co. Washington, D.C.	Technical Integration and Evaluation
General Electric-Apollo Support Dept., Daytona Beach, Fla.	Apollo Checkout and Reliability
North American Rockwell Corp. Space Div , Downey, Calif.	Spacecraft Command and Service Modules
Grumman Aircraft Engineering Corp., Bethpage, N.Y.	Lunar Module
Massachusetts Institute of Technology, Cambridge, Mass.	Guidance & Navigation (Technical Management)
General Motors Corp., AC Electronics Div., Milwaukee	Guidance & Navigation (Manufacturing)

TRW Systems Inc. Redondo Beach, Calif.	Trajectory Analysis
Avco Corp Space Systems Div. Lowell, Mass.	Heat Shield Ablative Material
North American Rockwell Corp. Rocketdyne Div., Canoga Park, Ca	J-2 Engines, F-1 Engines
The Boeing Co. New Orleans	First Stages (SIC) of Saturn V Flight Vehicles, Saturn V Systems Engineering and Integration Ground Support Equipment
North American Rockwell Corp. Space Div. Seal Beach, Calif.	Development and Production of Saturn V Second Stage (S-II)
McDonnell Douglas Astronautics Co. Huntington Beach, Calif.	Development and Production of Saturn V Third Stage (S-IVB)
International Business Machines Federal Systems Div. Huntsville, Ala.	Instrument Unit (Prime Contractor)
Bendix Corp. Navigation and Control Div. Teterboro, N. J.	Guidance Components for Instrument Unit (Including ST-124M Stabilized Platform)
Trans World Airlines, Inc.	Installation Support, KSC
Federal Electric Corp.	Communications and Instrumentation Support, KSC
Bendix Field Engineering Corp.	Launch Operations/Complex Support, KSC
Catalytic-Dow	Facilities Engineering and Modifications, KSC
ILC Industries Dover, Del.	Space Suits
Radio Corp. of America Van Nuys, Calif.	110A Computer - Saturn Checkout
Sanders Associates Nashua, N. H.	Operational Display Systems Saturn
Brown Engineering Huntsville, Ala.	Discrete Controls
Ingalls Iron Works Birmingham, Ala.	Mobile Launchers (structural work)
Smith/Ernst (Joint Venture) Tampa, Fla. Washington, D.C.	Electrical Mechanical Portion of MLs
Power Shovel: Inc. Marion, Ohio	Crawler-Transporter
Hayes International Birmingham, Ala.	Mobile Launcher Service Arms

APOLLO 9 GLOSSARY

Ablating Materials— Special heat-dissipating materials on the surface of a spacecraft that can be sacrificed (carried away, vaporized) during reentry.

Abort— The unscheduled intentional termination of a mission prior to its completion.

Accelerometer— An instrument to sense accelerative forces and convert them into corresponding electrical quantities usually for controlling, measuring, indicating or recording purposes.

Adapter Skirt— A flange or extension of a stage or section that provides a ready means of fitting another stage or section to it.

Apogee— The point at which a Moon or artificial satellite in its orbit is farthest from Earth.

Attitude— The position of an aerospace vehicle as determined by the inclination of its axes to some frame of reference; for Apollo, an inertial, space-fixed reference is used.

Burnout— The point when combustion ceases in a rocket engine.

Canard— A short, stubby wing-like element affixed to the launch escape tower to provide CM blunt end forward aerodynamic capture during an abort.

Celestial Guidance— The guidance of a vehicle by reference to celestial bodies.

Celestial Mechanics— The science that deals primarily with the effect of force as an agent in determining the orbital paths of celestial bodies.

Closed Loop— Automatic control units linked together with a process to form an endless chain.

Deboost— A retrograde maneuver which lowers either perigee or apogee of an orbiting spacecraft. Not to be confused with deorbit.

Delta V— Velocity change

Digital Computer— A computer in which quantities are represented numerically and which can be used to solve complex problems.

Down-Link— The part of a communication system that receives, processes and displays data from a spacecraft.

Ephemeris— Orbital measurements (apogee, perigee, inclination, period, etc.) of one celestial body in relation to another at given times. In spaceflight, the orbital measurements of a spacecraft relative to the celestial body about which it orbited.

Explosive Bolts— Bolts destroyed or severed by a surrounding explosive charge which can be activated by an electrical impulse.

Fairing— A piece, part or structure having a smooth, streamlined outline, used to cover a non-streamlined object or to smooth a junction.

Flight Control System — A system that serves to maintain attitude stability and control during flight.

Fuel Cell — An electrochemical generator in which the chemical energy from the reaction of oxygen and a fuel is converted directly into electricity.

G or G Force— Force exerted upon an object by gravity or by reaction to acceleration or deceleration, as in a change of direction: one G is the measure of the gravitational pull required to accelerate a body at the rate of about 32.16 feet-per-second.

Gimbaled Motor— A rocket motor mounted on gimbal; i.e. on a contrivance having two mutually perpendicular axes of rotation, so as to obtain pitching and yawing correction moments.

Guidance System— A system which measures and evaluates flight information, correlates this with target data, converts the result into the conditions necessary to achieve the desired flight path, and communicates this data in the form of commands to the flight control system.

Inertial Guidance — Guidance by means of the measurement and integration of acceleration from onboard the spacecraft. A sophisticated automatic navigation system using gyroscopic devices, accelerometers etc., for high-speed vehicles. It absorbs and interprets such data as speed, position, etc., and automatically adjusts the vehicle to a pre-determined flight path. Essentially, it knows where it's going and where it is by knowing where it came from and how it got there. It does not give out any radio frequency signal so it cannot be detected by radar or jammed.

Injection— The process of boosting a spacecraft into a calculated trajectory.

Insertion— The process of boosting a spacecraft into an orbit around the Earth or other celestial bodies.

Multiplexing— The simultaneous transmission of two or more signals within a single channel. The three basic methods of multiplexing involve the separation of signals by time division, frequency division and phase division.

Optical Navigation— Navigation by sight, as opposed to inertial methods, using stars or other visible objects as reference.

Oxidizer— In a rocket propellant, a substance such as liquid oxygen or nitrogen tetroxide which supports combustion of the fuel.

Penumbra— Semi-dark portion of a shadow in which light is partly cut off, e.g. surface of Moon or Earth away from Sun. (See umbra.)

Perigee— Point at which a Moon or an artificial satellite in its orbit is closest to the Earth.

Pitch— The angular displacement of a space vehicle about its lateral axis (Y).

Reentry— The return of a spacecraft that reenters the atmosphere after flight above it.

Retrorocket— A rocket that gives thrust in a direction opposite to the direction of the object's motion.

Roll— The angular displacement of a space vehicle about its longitudinal (X) axis.

S-Band— A radio-frequency band of 1,550 to 5,200 megahertz.

Sidereal— Adjective relating to measurement of time, position or angle in relation to the celestial sphere and the vernal equinox.

State vector— Ground generated spacecraft position, velocity and timing information uplinked to the spacecraft computer for crew use as a navigational reference.

Telemetering— A system for taking measurements within an aerospace vehicle in flight and transmitting them by radio to a ground station.

Terminator— Separation line between lighted and dark portions of celestial body which is not self luminous.

Ullage— The volume in a closed tank or container that is not occupied by the stored liquid; the ratio of this volume to the total volume of the tank; also an acceleration to force propellants into the engine pump intake lines before ignition.

Umbra— Darkest part of a shadow in which light is completely absent, e.g. surface of Moon or Earth away from Sun.

Update pad— Information on spacecraft attitudes, thrust values, event times, navigational data, etc., voiced up to the crew in standard formats according to the purpose, e.g. maneuver update, navigation check, landmark tracking, entry update, etc.

Up-Link Data— Information fed by radio signal from the ground to a spacecraft.

Yaw— Angular displacement of a space vehicle about its vertical (Z) axis.

APOLLO 9 ACRONYMS AND ABBREVIATIONS

(Note: This list makes no attempt to include all Apollo program acronyms and abbreviations, but several are listed that will be encountered frequently in the Apollo 9 mission. Where pronounced as words in air-to-ground transmissions, acronyms are phonetically shown in parentheses. Otherwise, abbreviations are sounded out by letter.)

AGS	(Aggs)	Abort Guidance System (LM)
AK		Apogee kick
APS	(Apps)	Ascent Propulsion System
BMAG	(Bee-mag)	Body mounted attitude gyro
CDH		Constant delta height
CMC		Command Module Computer
COI		Contingency orbit insertion
CSI		Concentric sequence initiate
DAP	(Dapp)	Digital autopilot
DEDA	(Dee-da)	Data Entry and Display Assembly (LM AGS)

DFI		Development flight instrumentation
DPS	(Dips)	Descent propulsion system
DSKY	(Diskey)	Display and keyboard
FDAI		Flight director attitude indicator
FITH	(Fith)	Fire in the hole (LM ascent staging)
FTP		Fixed throttle point
HGA		High-gain antenna
IMU		Inertial measurement unit
IRIG	(Ear-ig)	Inertial rate integrating gyro
MCC		Mission Control Center
MC&W		Master caution and warning
MTVC		Manual thrust vector control
NCC		Combined corrective maneuver
NSR		Co-elliptical maneuver
PIPA	(Pippa)	Pulse integrating pendulous accelerometer
PLSS	(Pliss)	Portable life support system
PUGS	(pugs)	Propellant utilization and gauging system
REFSMMAT	(Refsmat)	Reference to stable member matrix
RHC		Rotation hand controller
RTC		Real-time command
SCS		Stabilization and control system
SLA	(Slah)	Spacecraft LM adapter
SPS		Service propulsion system
THC		Thrust hand controller
TPF		Terminal phase finalization
TPI		Terminal phase initiate
TVC		Thrust vector control

CONVERSION FACTORS

Multiply	By	To Obtain
Distance:		
feet	0.3048	meters
meters	3.281	feet
kilometers	3281	feet
kilometers	0.6214	statute miles
statute miles	1.609	kilometers
nautical miles	1.852	kilometers
nautical miles	1.1508	statute miles
statute miles	0.86898	nautical miles
statute mile	1760	yards
Velocity:		
feet/sec	0.3048	meters/sec
meters/sec	3.281	feet/sec
meters/sec	2.237	statute mph
feet/sec	0.6818	statute miles/hr
feet/sec	0.5925	nautical miles/hr
statute miles/hr	1.609	km/hr
nautical miles/hr (knots)	1.852	km/hr
km/hr	0.6214	statute miles/hr
Liquid measure weight		
gallons	3.785	liters
liters	0.2642	gallons
pounds	0.4536	kilograms
kilograms	2.205	pounds
Volume:		
cubic feet	0.02832	cubic meters
Pressure:		
pounds/sq inch	70.31	grams/sq cm

MISSION OPERATION REPORT

APOLLO 9 (AS-504) MISSION

OFFICE OF MANNED SPACE FLIGHT
Prepared by: Apollo Program Office-MAO

FOR INTERNAL USE ONLY - February 18th 1969
M-932-69-09

MEMORANDUM

To: A/Acting Administrator

From: MA/Apollo Program Director

Subject: Apollo 9 Mission (AS-504)

No earlier than 28 February 1969, we plan to launch the next Apollo/Saturn V mission, Apollo 9. This will be the second manned Saturn V flight, the third flight of a manned Apollo Command/Service Module, and the first flight of a manned Lunar Module.

The purpose of this mission is to demonstrate crew/space vehicle/mission support facilities performance during a manned Saturn V mission with CSM and LM, demonstrate LM/crew performance, demonstrate performance of nominal and selected backup Lunar Orbit Rendezvous mission activities, and assess CSM/LM consumables.

The launch will be the fourth Saturn V from Launch Complex 39A at Kennedy Space Center. The launch window opens at 11:00 EST, closes at 14:00 EST and is determined by lighting and ground station coverage requirements for rendezvous.

The nominal mission will include: ascent to orbit, CSM transposition and docking, CSM/LM separation from the S-IVB, five docked SPS burns, two unmanned S-IVB restarts, LM systems evaluation, a docked DPS burn, EVA, LM active rendezvous, an unmanned APS burn to propellant depletion, CSM solo activities and three SPS burns, CM reentry, and recovery in the Atlantic.

Sam - Phillips
Lt. General, USAF
Apollo Program Director

George E. Mueller
Associate Administrator for Manned Space Flight

FOREWORD

MISSION OPERATION REPORTS are published expressly for the use of NASA Senior Management, as required by the Administrator in NASA Instruction 6-2-10, dated August 15, 1963. The purpose of these reports is to provide NASA Senior Management with timely, complete, and definitive information on flight mission plans, and to establish official mission objectives which provide the basis for assessment of mission accomplishment.

Initial reports are prepared and issued for each flight project just prior to launch. Following launch, updating reports for each mission are issued to keep General Management currently informed of definitive mission results as provided in NASA Instruction 6-2-10.

Because of their sometimes highly technical orientation, distribution of these reports is limited to personnel having program-project management responsibilities. The Office of Public Affairs publishes a comprehensive series of pre-launch and post launch reports on NASA flight missions, which are available for general distribution.

APOLLO 9 MISSION OPERATION REPORT

The Apollo 9 Mission Operation Report is published in two volumes I, The Mission Operation Report (MOR) and II, the Mission Operation Report Supplement.

This format was designed to provide a mission-oriented document in the MOR with only a very brief description of the space vehicle and support facilities. The MOR Supplement is a reference document with a more comprehensive description of the space vehicle, launch complex, and mission monitoring, support, and control facilities.

GENERAL

The goal of the Apollo Program is to enhance the manned space flight capability of the United States by developing, through logical and orderly evolution, the ability to land men on the moon and return them safely to earth.

To accomplish the goal of lunar landing and return in this decade, the Apollo Program has focused on the development of a highly reliable launch vehicle and spacecraft system. This has been done through a logical sequence of Apollo missions designed to qualify the flight hardware, ground support systems, and operational personnel in the most effective manner.

The Apollo 9 mission is the third manned flight of the Apollo Command and Service Module (CSM), the first manned flight of the Lunar Module (LM), and the second manned Saturn V mission. The mission is designed to test the space vehicle, mission support facilities, and crew with a complete Apollo spacecraft (CSM and LM) in earth orbit.

PROGRAM DEVELOPMENT

The first Saturn vehicle was successfully flown on 27 October 1961 to initiate operations of the Saturn I Program.

A total of 10 Saturn I vehicles (SA-1 to SA-10) were successfully flight tested to provide information on the integration of launch vehicle and spacecraft and to provide operational experience with large multi-engine booster stages (S-1, S-IV).

The next generation of vehicles, developed under the Saturn IB Program, featured an uprated first stage (S-IB) and a more powerful new second stage (S-IVB). The first Saturn IB was launched on 26 February 1966. The first three Saturn IB missions (AS-201, AS-203, and AS-202) successfully tested the performance of the launch vehicle and spacecraft combination, separation of the stages, behavior of liquid hydrogen in a weightless environment, performance of the Command Module heat shield at low earth orbital entry conditions, and recovery operations.

The planned fourth Saturn IB mission (AS-204) scheduled for early 1967 was intended to be the first manned Apollo flight. This mission was not flown because of a spacecraft fire, during a manned pre-launch test, that took the lives of the prime flight crew and severely damaged the spacecraft. The SA-204 launch vehicle was later assigned to the Apollo 5 mission.

The Apollo 4 mission was successfully executed on 9 November 1967. This mission initiated the use of the Saturn V launch vehicle (SA-501) and required an orbital restart of the S-IVB third stage. The spacecraft for this mission consisted of an unmanned Command and Service Module (CSM) and a Lunar Module Test Article (LTA). The CSM Service Propulsion System (SPS) was exercised, including restart, and the Command Module Block II heat shield was subjected to the combination of high heat load, high heat rate, and aerodynamic load representative of lunar return entry.

The Apollo 5 mission was successfully launched and completed on 22 January 1968. This was the fourth mission utilizing Saturn IB vehicles (SA-204). This flight provided for unmanned orbital testing of the Lunar Module (LM-1). The LM structure, staging, and proper operation of the Lunar Module Ascent Propulsion System (APS) and Descent Propulsion System (DPS), including restart, were verified. Satisfactory performance of the S-IVB/Instrument Unit (IU) in orbit was also demonstrated.

The Apollo 6 mission (second unmanned Saturn V) was successfully launched on 4 April 1968. Some flight anomalies encountered included oscillation reflecting propulsion structural longitudinal coupling, an imperfection in the Spacecraft LM Adapter (SLA) structural integrity, and certain malfunctions of the J-2 engines in the S-II and S-IVB stages. The spacecraft flew the planned trajectory, but pre-planned high velocity reentry conditions were not achieved. A majority of the mission objectives for Apollo 6 were accomplished.

The Apollo 7 mission (first manned Apollo) was successfully launched on 11 October 1968. This was the fifth and last planned Apollo mission utilizing Saturn IB launch vehicles (SA-205). The eleven-day mission provided the first orbital tests of the Block II Command and Service Module. All Primary Mission Objectives were successfully accomplished. In addition, all planned Detailed Test Objectives, plus three that were not originally scheduled, were satisfactorily accomplished.

The Apollo 8 mission was successfully launched on 21 December and completed on 27 December 1968. This was the first manned flight of the Saturn V launch vehicle and the first manned flight to the vicinity of the moon. All Primary Mission Objectives were successfully accomplished. In addition, all Detailed Test Objectives plus four that were not originally scheduled, were successfully accomplished. Ten orbits of the moon were successfully performed with the last eight circular at an altitude of 60 nautical miles. TV and film photographic coverage was successfully carried out, with telecasts to the public being made in real time.

THE APOLLO 9 MISSION

Apollo 9 (AS-504) will be the fourth Saturn V Mission and the first manned flight of the Lunar Module (LM). Mission duration is open-ended and currently planned for approximately ten days (239 hours). This CSM/LM Operations Mission is designed to achieve an evaluation of a manned LM and to demonstrate the compatibility of the CSM and LM to perform combined operations typical of lunar missions.

The mission has been divided into six activity periods over the ten-day mission duration. The first activity period will consist of launch and ascent to orbit, CSM transposition and docking with the LM/IU/S-IVB, separation of the CSM/LM from the IU/S-IVB, two unmanned S-IVB engine restarts, and one docked CSM/LM SPS burn. The second activity period will be taken up primarily by three more SPS burns. The third activity period will see the first LM operation including extensive LM systems checkout and the first docked CSM/LM DPS burn followed by the fifth SPS burn. The fourth activity period will be devoted to extravehicular activity (EVA) including a transfer from the LM to the CSM and return, and live TV of the CSM and LM by the EVA astronaut. The fifth activity period will complete LM activities with two DPS burns, LM descent stage jettison, LM-active rendezvous including the first APS burn, and finally APS burn to propellant depletion. The sixth activity period will be devoted to CSM solo operations including three SPS burns, navigation exercises, a multispectral terrain photography scientific experiment, CM reentry, and recovery.

NASA OMSF PRIMARY MISSION OBJECTIVES
FOR APOLLO 9

PRIMARY OBJECTIVES

Demonstrate crew/space vehicle/mission support facilities performance during a manned Saturn V mission with CSM and LM.

Demonstrate LM/Crew performance.

Demonstrate performance of nominal and selected backup Lunar Orbit Rendezvous (LOR) mission activities including:

- Transposition, docking, LM withdrawal
- Inter-vehicular crew transfer
- Extra-vehicular capability
- SPS and DPS burns
- LM active rendezvous and docking

CSM/LM consumables assessment.

Sam C. Phillips
Lt. General, USAF
Apollo Program Director

George E. Mueller
Associate Administrator for
Manned Space Flight

Date: 14 FEB 69

DETAILED TEST OBJECTIVES

Mandatory and Principal Detailed Test Objectives (DTO's) amplify and define more explicitly those basic tests, measurements, and evaluations that are planned to achieve the Primary Objectives of the Apollo 9 Mission.

Launch Vehicle

Demonstrate S-IVB/IU attitude control capability during transposition, docking, and LM ejection (T, D, & E) maneuver.

Spacecraft

Perform LM-active rendezvous (20.27).

Determine DPS duration effects and primary propulsion/vehicle interactions (13.12).

Verify satisfactory performance of passive thermal subsystem (17.17).

Demonstrate LM structural integrity (17.18).

Perform DPS burn including throttling, docked; and a short duration DPS burn, undocked (11.6).

Perform long duration APS burns (13.11).

Demonstrate Environmental Control System (ECS) performance during all LM activities (14.0).

Obtain temperature data on deployed landing gear resulting from DPS operation (17.9).

Determine Electrical Power System (EPS) performance, primary and backup (15.3).

Operate landing radar during DPS burns (16.7).

Perform Abort Guidance System (AGS)/Central Electronics System (CES) controlled DPS burn (12.4).

Perform Primary Guidance, Navigation, and Control System (PGNCS)/Digital Auto Pilot (DAP) controlled long duration APS burn (11.14).

Demonstrate RCS control of LM using manual and automatic PGNCS (11.7).

Demonstrate S-band and VHF communication compatibility (20.22).

Demonstrate RCS control of LM using manual and automatic AGS/CES (12.3).

Demonstrate CSM attitude control, docked, during SPS burn (1.23).

Demonstrate LM-Active docking (20.28).

Demonstrate LM ejection from SLA with CSM (20.25).

Demonstrate CSM - active docking (20.24).

Demonstrate LM-CSM undocking (20.26).

Verify Inertial Measurement Unit (IMU) performance (11.10).

Demonstrate Guidance, Navigation, and Control System (GNCS)/Manual Thrust Vector Control (MTVC) takeover (2.9).

Demonstrate LM rendezvous radar performance (16.4).

Demonstrate LM/Manned Space Flight Network (MSFN) S-band communications capability (20.21).

Demonstrate IVT (20.34).

Demonstrate AGS calibration and obtain performance data in flight (12.2).

Perform LM IMU alignment (11.5).

Perform LM jettison (20.29).

Obtain data on Reaction Control System (RCS) plume impingement and corona effect on rendezvous radar performance (16.19).

Demonstrate support facilities performance during earth orbital missions (20.31).

Perform IMU alignment and daylight star visibility check, docked (1.25).

Prepare for CSM-active rendezvous with LM (20.33).

Perform IMU alignment with sextant (SXT), docked (1.24).

Perform landing radar self-test (16.6).

Extravehicular Activity (20.35).

SECONDARY OBJECTIVES
Apollo secondary objectives are established by the development centers to provide additional engineering or scientific data.

Launch Vehicle

Verify S-IVB restart capability.

Verify J-2 engine modifications.

Confirm J-2 engine environment in S-II stage.

Confirm launch vehicle longitudinal oscillation environment during S-IC stage burn period.

Demonstrate O_2H_2 burner repressurization system operation.

Demonstrate S-IVB propellant dump and safing.

Verify that modifications incorporated in the S-IC stage suppress low-frequency longitudinal oscillations.

Demonstrate 80-minute restart capability.

Demonstrate dual repressurization capability.

Demonstrate O_2H_2 burner restart capability.

Verify the onboard Command and Communications System (CCS)/ground system interface and operation in the space environment.

Spacecraft

Obtain exhaust effects data from Launch Escape Tower (LET), S-II retro, and SM RCS on CSM (7.29).

Evaluate crew performance of all tasks (20.32).

Perform navigation by landmark tracking (1.26).

Perform APS burn-to-depletion, unmanned (13.10).

Obtain data on DPS plume effects on visibility (20.37).

Perform CSM/LM electromagnetic compatibility test (20.120).

LAUNCH COUNTDOWN AND TURNAROUND CAPABILITY AS-504

COUNTDOWN

Countdown for the Apollo 9 mission will begin with a pre-count period starting at T-130.5 hours during which launch vehicle (LV) and spacecraft preparations will take place independently. Coordinated space vehicle (SV) launch countdown activities begin at T-28 hours. Table I shows the significant launch countdown events.

SCRUB/TURNAROUND

Scrub/turnaround times are based upon the amount of work required to return the space vehicle to a safe condition and to complete the recycle activities necessary to resume launch countdown for a subsequent launch attempt. Planning guidelines for the various scrub/turnaround plans are based upon no serial time for repairs or holds, or for systems retesting resulting from repairs; performing tasks necessary to attain launch with the same degree of confidence as for the first launch attempt; and, not requiring unloading of hypergolic propellants and RP-I from the SV.

TURNAROUND CONDITIONS VS. TIME*

Scrub can occur at any point in the countdown when weather, launch support facilities or SV conditions warrant. For a hold that results in a scrub prior to T-22 minutes, turnaround procedures are initiated from the point of hold. Should a hold occur from T-22 minutes (S-II start bottle chilldown) to T-16.2 seconds (S-IC forward umbilical disconnect), then a recycle to T-24 minutes, hold, or scrub is possible under the conditions stated in the Launch Mission Rules. An automatic or manual cutoff after T-16.2 seconds will result in a scrub. For planning purposes, four primary cases are identified to implement the required turnaround activities for subsequent launch attempt following a countdown scrub.

Post-LV Cryogenic Load (with Fuel Cell Cryogenic Reservicing)

Turnaround time is 70 hours, 15 minutes, consisting of 42 hours, 15 minutes for recycle time and 28 hours for countdown time. Turnaround time is based upon; scrub occurs between 16.2 seconds and 8.9 seconds during original countdown; reservicing of the CSM or LM water systems not required; all SV ordnance except Range Safety Destruct Safe and Arm (S&A) units remain connected; access into LM cabin not required. The time required for this turnaround results from flight crew egress; LV cryogenic unloading; LV ordnance operations and battery removal; LM supercritical helium (SHe) reservicing; CSM cryogenic reservicing; CSM battery removal and installation, and countdown resumption at T-28 hours.

The information in this section is based on KSC Scrub /Turnaround Plan for Apollo 9, dated 23 January 1969, and is subject to change.

TABLE I   LAUNCH COUNTDOWN SEQUENCE OF EVENTS

COUNTDOWN

HRS: MIN: SEC	EVENT
28:00:00	Start Countdown Clock
21:30:00 R	Start LM Stowage and Cabin Closeout
24:30:00 R	Start LM Systems Checks
19:30:00	Start SV EDS Test
13:30:00	Remove LM Platform
12:15:00	SLA Closeout Complete
11:30:00 R	CSM Pre-ingress Operations
09:30:00	CSM RF Voice Checks
09:15:00 R	LV Closeout Complete
09:00:00	Six-Hour built-in Hold
09:00:00	Clear Blast Danger Area
08:30:00	Start MSS Move to Park Site
08:15:00	Start LV Propellant Loading
03:38:00	Start LV Propellant Replenishing
02:40:00	Start Flight Crew Ingress
02:10:00	Flight Crew Ingress Completed
01:55:00	Start MCC/CSM Command Checks
01:15:00	Release Final Jimsphere
00:43:00	Retract SA-9 to 12° Park Position
00:42:00	Arm LES
00:30:00	LM to Internal Power
00:15:00	CSM to Internal Power
00:05:00	SA-9 Fully Retracted
00:03:07	Start Terminal Count Sequencer
00:00:08.9	Ignition Command
00:00:00	First Motion

R = Reference times. These times are approximately when the event will occur and could be adjusted prio
to start of countdown clock.

Post-LV Cryogenic Load (No Fuel Cell Cryogenic Reservicing)

Turnaround time is 38 hours, 30 minutes, consisting of 29 hours, 30 minutes for recycle time and 9 hours fo
countdown time. Turnaround time is based upon: scrub occurs between 16.2 seconds and 8.9 seconds durin
original countdown; the Range Safety Destruct S&A units will remain connected; the CSM batteries do no
require replacement; reservicing of the CSM or LM water systems not required; access to LM cabin no
required. The time for this turnaround results from flight crew egress, LV cryogenic unloading, LM SH
reservicing, LV loading preparations, and countdown resumption at T-9 hours.

Pre-LV Cryogenic Load (with Fuel Cell Cryogenic Reservicing)

Turnaround time is 53 hours, 30 minutes, consisting of 44 hours, 30 minutes for recycle time and 9 hours fo
countdown time. Turnaround time is based upon: scrub occurs at T-8 hours of the original countdown; th
Range Safety Destruct S&A units remain connected; the required S-II servo-actuator inspection is wavere
reservicing of CSM or LM water systems not required; and access into LM cabin not required. The tim
required for this turnaround results from CSM cryogenic reservicing, CSM battery removal and installatio
LM SHe reservicing, and countdown resumption at T-9 hours.

Pre-LV Cryo Load (No Fuel Cell Cryo Reservicing)

The capability for a one-day turnaround exists at T-8 hours of the countdown. This capability provides for a launch attempt at the opening of the next launch window. Turnaround time is 32 hours, consisting of 23 hours for recycle time and 9 hours for countdown time. Turnaround time is based upon: scrub occurred at T-8 hours of the countdown; the required S-II servo-actuator inspection is wavered; the Range Safety Destruct S&A units remain connected; CSM batteries do not require replacement; reservicing of CSM or LM water systems not required; access into LM cabin not required. The time required for this turnaround results from LM SHe reservicing and countdown resumption at T-9 hours.

DETAILED FLIGHT MISSION DESCRIPTION

NOMINAL MISSION

The Apollo 9 Mission Plan is divided into six activity periods which span eleven work days over eleven calendar days. The relationships among these periods and days, and the major scheduled activities are shown to scale in Figure 1. Each work day will terminate with completion of a crew sleep period.

A summary profile of the Apollo 9 mission is shown in Figure 2 and a detailed summary of the Apollo 9 Flight Plan is given in Figure 3.

The sequence of events for the Apollo 9 mission is given in Table 2. Launch vehicle (LV) time base (TB) notations are also included. Time bases may be defined as precise initial points upon which succeeding critical preprogrammed activities or functions may be based. The TB's noted in Table 2 are for a nominal mission and presuppose nominal LV performance. However, should the launch vehicle stages produce non-nominal performance, the launch vehicle computer will recompute the subsequent TB's and associated burns to correct LV performance to mission rules.

First Period of Activity (Figure 4)

The Apollo 9 mission will be launched from Kennedy Space Center, Launch Complex 39, Pad A, on a flight azimuth of 72° . The launch window opens at 1100 EST, closes at 1400 EST, and is determined by lighting and ground station coverage requirements for CSM/LM rendezvous.

The Apollo 9 mission will begin with full S-IC and S-II launch vehicle stage burns and partial burn of the S-IVB stage to insert the S-IVB, Instrument Unit (IU) Lunar Module (LM), and Command/Service Module (CSM) into a 103 nautical mile near-circular orbit.

Immediately after insertion, the crew will begin a series of CSM/S-IVB orbital operations. This activity is to configure the CSM for orbital operations, prepare for transposition, docking and ejection, and to evaluate the operations required to verify that the S-IVB/ IU/LM/CSM would be ready for trans-lunar injection (TLI) on a lunar mission. The actual TLI burn will not be performed on this mission. Venting of S-IVB after insertion will raise apogee to approximately 112 nautical miles and perigee to approximately 109 nautical miles after about 2 hours, 30 minutes GET (ground elapsed time). At about 2 hours, 40 minutes GET, the CSM will separate from the S-IVB/IU/LM and a visual inspection of the S-IVB/IU/LM will be performed from the CSM prior to docking the CSM to the LM/IU/S-IVB. Immediately after docking, the LM pressurization will begin and, upon full pressure verification, LM ejection from the SLA will be initiated. Following ejection, a small Service Module Reaction Control System (SM RCS) burn will be executed to separate the CSM/LM to a safe distance prior to the first unmanned S-IVB restart (second burn). After the second S-IVB burn, a small docked-CSM/LM Service Propulsion System (SPS) burn will be performed to raise apogee of the CSM/LM orbit to 128 nautical miles. The S-IVB and SPS burns will be spaced to take advantage of the MSFN ground coverage before the end of the first activity period. A little after six hours GET, the S-IVB will be ignited a third time producing a sufficient increase in velocity to go into solar orbit.

APOLLO 9 MISSION TIME AND EVENT CORRELATION

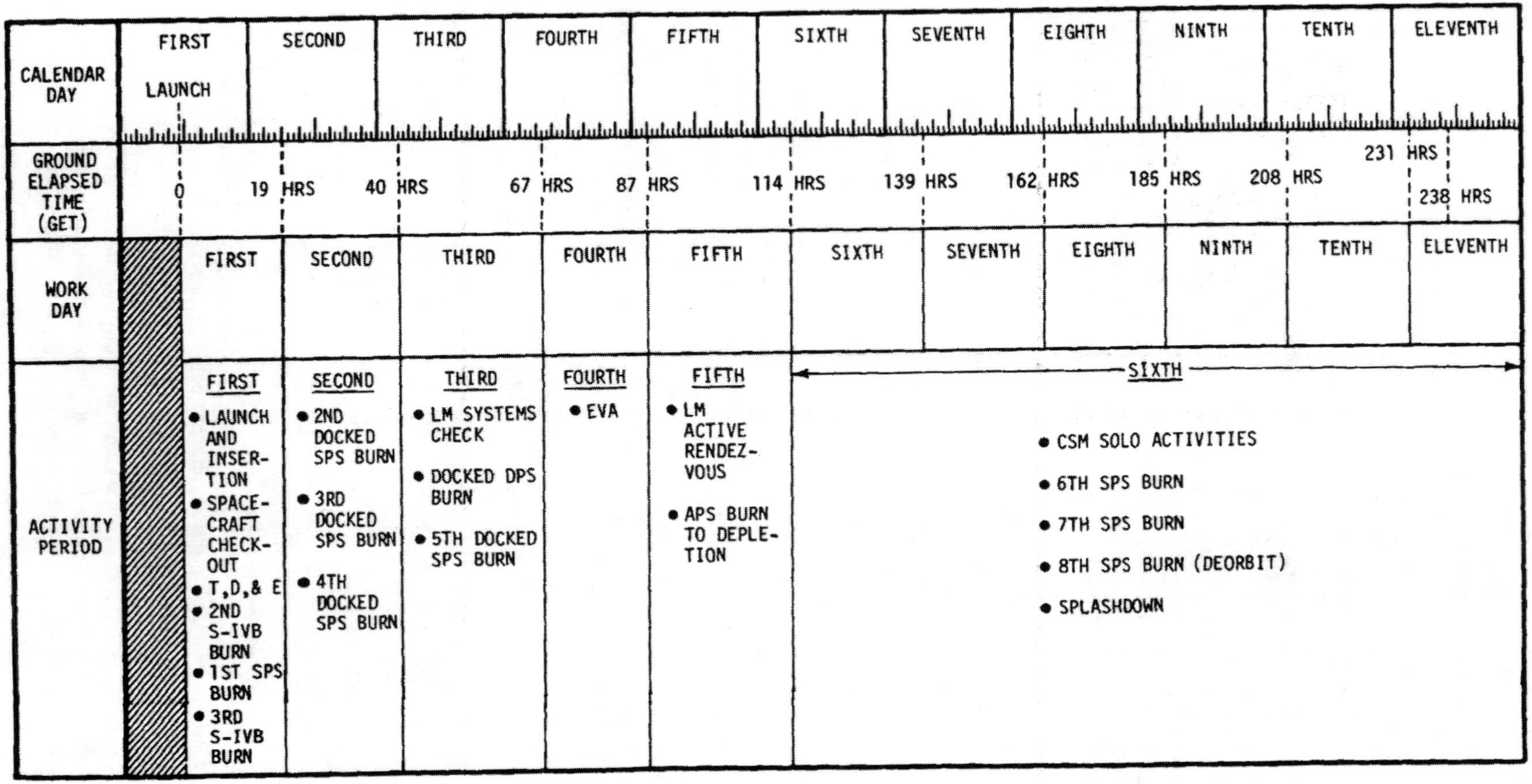

CALENDAR DAY	GROUND ELAPSED TIME (GET)	WORK DAY	ACTIVITY PERIOD	
FIRST	0			LAUNCH
FIRST		FIRST	FIRST	• LAUNCH AND INSERTION • SPACECRAFT CHECKOUT • T, D, & E • 1ST SPS BURN • S-IVB 2ND BURN • 3RD S-IVB BURN
SECOND	19 HRS	SECOND	SECOND	• 2ND DOCKED SPS BURN • 3RD DOCKED SPS BURN • 4TH DOCKED SPS BURN
THIRD	40 HRS	THIRD	THIRD	• LM SYSTEMS CHECK • DOCKED DPS BURN • 5TH DOCKED SPS BURN
FOURTH	67 HRS	FOURTH	FOURTH	• EVA
FIFTH	87 HRS	FIFTH	FIFTH	• LM ACTIVE RENDEZVOUS • APS BURN TO DEPLETION
SIXTH	114 HRS	SIXTH	SIXTH	• CSM SOLO ACTIVITIES • 6TH SPS BURN • 7TH SPS BURN • 8TH SPS BURN (DEORBIT) • SPLASHDOWN
SEVENTH	139 HRS	SEVENTH		
EIGHTH	162 HRS	EIGHTH		
NINTH	185 HRS	NINTH		
TENTH	208 HRS	TENTH		
ELEVENTH	231 HRS 238 HRS	ELEVENTH		

Fig. I

APOLLO 9 (AS-504) MISSION PROFILE

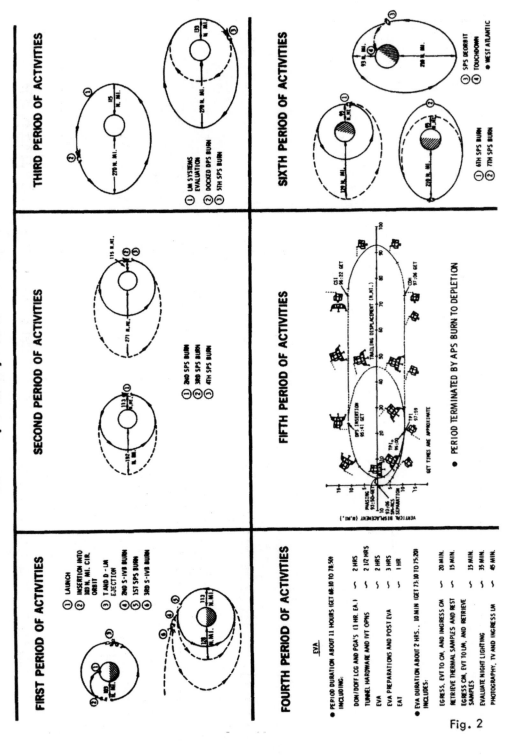

Fig. 2

APOLLO 9 SUMMARY FLIGHT PLAN

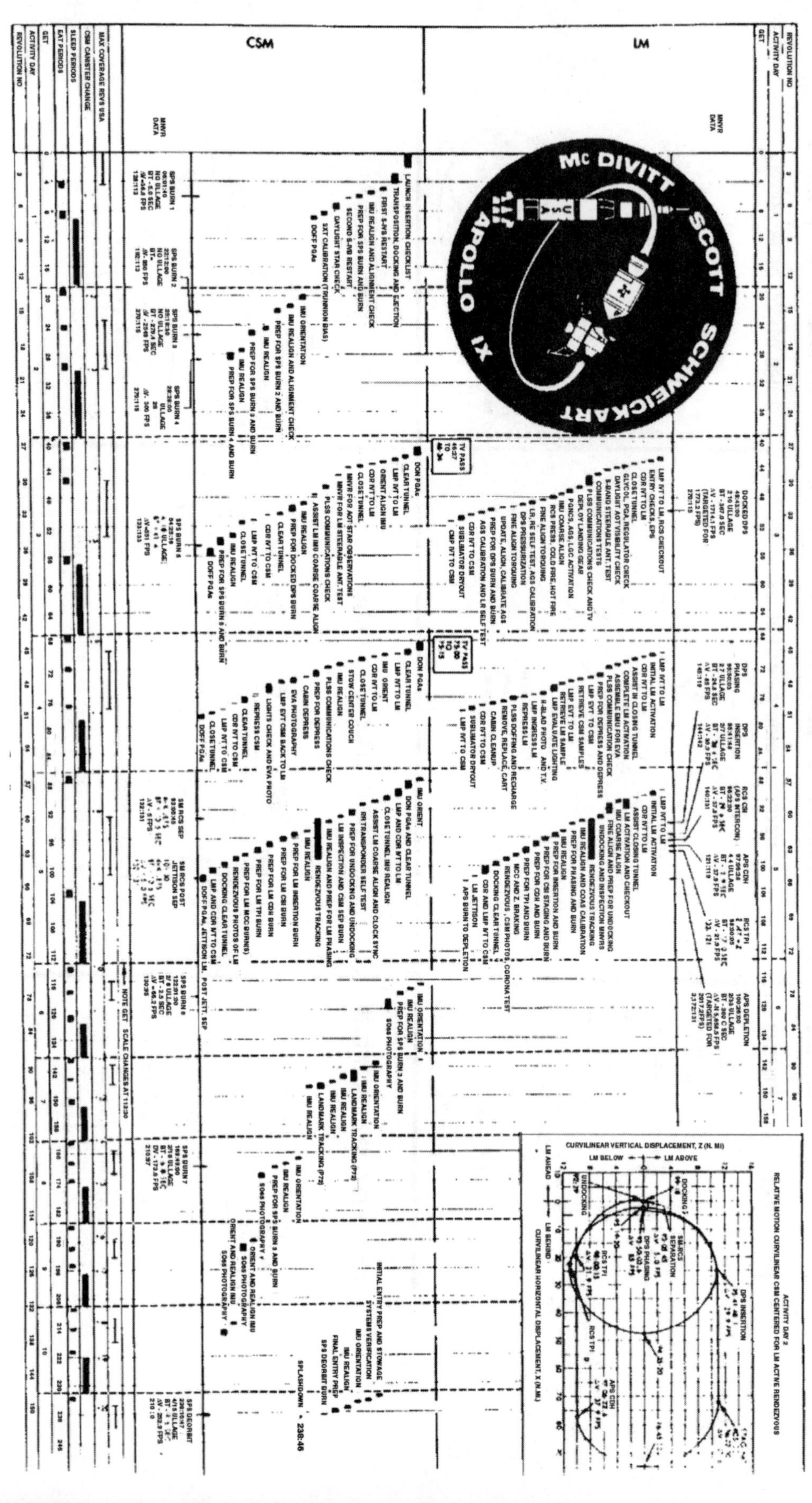

Fig. 3

TABLE 2 APOLLO 9 SEQUENCE OF EVENTS

*Ground Elapsed Time(GET):

HR:MIN:SEC*	EVENT
00:00:00	First Motion
00:00:00.391	Time base I
00:01:21	Maximum Dynamic Pressure
00:02:14	S-IC Center Engine Cutoff - TB2
00:02:40	S-IC Outboard Engine Cutoff - TB3
00:02:40	S-IC/S-II Separation
00:02:42	S-II Ignition
00:03:10	Jettison S-II Aft Interstage
00:03:16	Jettison Launch Escape Tower
00:08:51	S-II Engine Cutoff Command - TB4
00:08:52	S-II/S-IVB Separation
00:08:55	S-IVB Engine Ignition
00:10:49	S-IVB Engine Cutoff - TB5
00:10:59	Parking Orbit Insertion
02:33:49	Separation and Docking Maneuver Initiation
02:43:00	Spacecraft Separation
03:05:00	Spacecraft Docking (Approximately)
04:08:57	Spacecraft Final Separation
04:36:12	S-IVB Restart Preps. - TB6
04:45:50	S-IVB Re-ignition (2nd burn)
04:46:52	S-IVB Second Cutoff Signal - TB7
04:47:02	Intermediate Orbit Insertion
05:59:35	S-IVB Restart Preparations - TB8
06:01:40	SPS Burn I
06:07:13	S-IVB Re-ignition (3rd burn)
06:11:14	S-IVB Third Cutoff Signal - TB9
06:11:24	Escape Orbit Injection
06:12:44	Start LOX Dump
06:23:54	LOX Dump Cutoff
06:24:04	Start LH2 Dump
06:42:19	LH2 Dump Cutoff
22:12:00	SPS Burn 2
25:18:30	SPS Burn 3
28:28:00	SPS Burn 4
46:27:00	TV Transmission, LM Interior
49:42:00	Docked DPS Burn
54:25:19	SPS Burn 5
75:00:00	TV Transmission, CSM/LM Exterior by EVA LMP
92:39:00	Undocking
93:07:40	CSM/LM Separation
93:51:34	DPS Phasing
95:43:22	DPS Insertion
96:21:00	Concentric Sequence Initiation - RCS Burn
97:05:27	Constant Delta Height - APS Burn
98:00:10	Terminal Phase Initiation
99:13:00	CSM/LM Docking (Approximately)
101:58:00	APS Burn to Propellant Depletion
121:58:48	SPS Burn 6
169:47:54	SPS Burn 7
238:45:00	SPS Burn 8 (Deorbit)
238:59:47	Entry Interface (400,000 feet)
239:10:38	Drogue Chute Deployment (25,000 feet approx)
239:16:	Splashdown (Approximately)

*LV events based on MSFC LV Operational Trajectory, dated 31 January 1969.
SC events based on MSC SC Operational Trajectory, Revision 2, 20 February 7969.

APOLLO 9 LAUNCH DAY

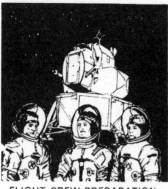

FLIGHT CREW PREPARATION

ORBITAL INSERTION

103 N. MILE ORBIT

SEPARATION

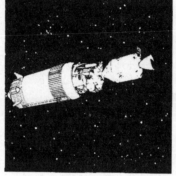

DOCKING

DOCKED SPS BURN

Fig 4

Second Period of Activity (Figure 5)

The second period of activity will be comprised of three docked-CSM/LM SPS burns. The first two burns in this period will be long duration, out-of-plane, and will adjust perigee to 115 nautical miles and apogee to 271 nautical miles. These burns will satisfy the CSM docked Digital Auto Pilot (DAP) stability margin test objective. These burns will also reduce the CSM weight to a level consistent with the SM RCS propellant requirements to deorbit or to effect a LM rescue if required during the fifth period LM active rendezvous. The burns will also provide nodal shift to assure proper lighting and rendezvous tracking. The third SPS burn in this period will be used to adjust phasing conditions for the LM- active rendezvous but will not change orbital parameters.

Third Period of Activity (Figure 6)

This period will be devoted primarily to checkout and performance evaluation of the LM systems including a docked Descent Propulsion System (DPS) engine burn of sufficient duration to evaluate performance of the LM Primary Guidance, Navigation, and Control System (PGNCS) digital auto-pilot and manual throttling of the DPS engine. Activities will begin with the inter-vehicular transfer (IVT) of the Commander (CDR), the Lunar Module Pilot (LMP), and equipment from the CSM to the LM. The LM systems will then be activated and checked out for the first time in the mission, commencing the systems performance evaluation. Following the docked DPS burn, the LM will be powered down and the CDR and LMP will return from the LM to the CSM via the IVT tunnel. A docked SPS burn will then circularize the orbit at 133 nautical miles and at the same time adjust the nodal position for the LM-active rendezvous. This burn will also purge helium from the SPS propellant feed system introduced as a result of the docked DPS burn. The first of two scheduled TV transmissions will be made during this period and will be of the LM interior.

APOLLO 9 SECOND DAY

LANDMARK TRACKING

PITCH MANEUVER

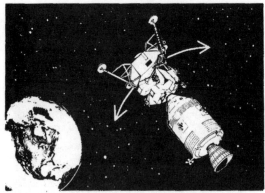

YAW-ROLL MANEUVER

HIGH APOGEE ORBITS

Fig 5

APOLLO 9 THIRD DAY

CREW TRANSFER

LM SYSTEM EVALUATION

Fig 6

APOLLO 9   FOURTH DAY

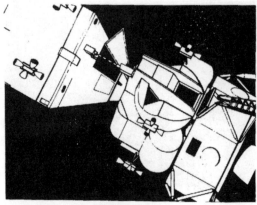

CAMERA

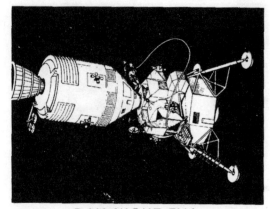

DAY-NIGHT EVA

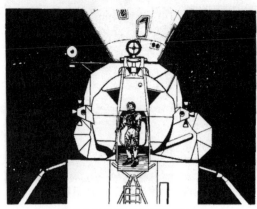

GOLDEN SLIPPERS

TV - TEXAS, FLORIDA

Fig 7

Fourth Period of Activity (Figure 7)

The fourth period will begin with IVT of the CDR and LMP to the LM, power-up and checkout of LM systems. Activities during this period will be in preparation for and accomplishment of the two-hour extravehicular activity (EVA) phase. The EVA phase will be initiated by the CDR and LMP performing IVT through the docking tunnel to the LM. The astronauts will next power up the LM and perform a systems check. All three astronauts will don their Pressure Garment Assemblies (PGA) subsequent to depressurizing the CM and LM. Upon depressurization, the LM forward batch and the CM side hatch will be opened. The LMP with the Portable Life Support System (PLSS) will leave the LM and attach a sequence camera to the LM handrail and another to the inside of the CM hatch. Both of these cameras will photograph the LMP's activities and will be remotely operated by the CDR and the Command Module Pilot (CMP).

Using the nominal transfer path, the LMP will transfer to the CM and partially enter through the side hatch after retrieving thermal samples from the SM and CM exterior. After a short rest period, the LMP will leave the CM and position himself on the LM porch. The LMP will remain outside of the LM and retrieve thermal samples, evaluate exterior lighting, photograph the LM and CSM, and operate the TV camera. At the end of these activities, the LMP will enter the LM through the forward hatch and subsequently transfer with the CDR to the CM through the docking tunnel.

Fifth Period of Activity (Figure 8)

The fifth period of activity will consist of the LM-active rendezvous and the unmanned long duration LM APB burn to propellant depletion. A schematic of the rendezvous is shown in Figure 3. The period will begin with

APOLLO 9   FIFTH DAY

VEHICLES UNDOCKED

LM - BURNS FOR RENDEZVOUS

MAXIMUM SEPARATION

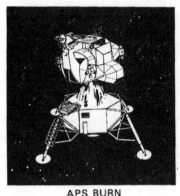
APS BURN

FORMATION FLYING AND DOCKING

LM JETTISON ASCENT BURN

Fig 8

the IVT of the LMP and CDR to the LM. The LM will be powered up and systems checked out prior to the LM being separated from the CSM for the first time in the mission. A short period of station-keeping will be performed prior to initiation of the phasing maneuver of the rendezvous. This phase of the rendezvous will begin with a short SM RCS burn directed in-plane and radially downward, placing the CSM and LM in small equi-period orbits (mini-football) from which a rendezvous abort can easily be made. This is the first rendezvous abort situation when the LM can rejoin the CSM.

Approximately one-half revolution later, an AGS (Abort Guidance System) controlled DPS phasing burn will be made in-plane and in a radially outward direction, placing the LM on an equi-period rendezvous trajectory (football) with the CSM. The purpose of performing the separation maneuver in this way is to expedite return to the CSM should the need arise. The second abort situation occurs at the first Terminal Phase Initiation (TPI_0) noted on Figure 3. At TPI_0 the LM can burn the RCS to effect an immediate rendezvous with the CSM or proceed to the nominal DPS insertion.

After approximately one and one-fourth revolutions, a PGNCS-controlled DPS insertion maneuver will be executed, placing the LM in a co-elliptic orbit above and behind the CSM. The remainder of this phase will be a Concentric Flight Plan (CFP) rendezvous sequence approaching the CSM from below and behind. The LM will be staged just prior to the Concentric Sequence Initiation (CSI) RCS burn. This burn is followed by the Ascent Propulsion System (APS) Constant Delta Height (CDH) burn and the RCS TPI burn. The rendezvous is terminated after a short station-keeping period. The total time for the rendezvous will be approximately six and one-half hours.

Following LM docking, the LM Ascent Stage will be prepared for an unmanned long duration APS burn. The LM crewmen will transfer to the CSM and the CSM will be separated from the LM. The long duration APS burn will then be performed on initiation by ground control and will terminate by propellant depletion.

APOLLO 9 SIXTH THRU NINTH DAYS

SERVICE PROPULSION BURNS

Fig 9

LANDMARK SIGHTINGS, PHOTOGRAPH
SPECIAL TESTS

Sixth Period of Activity (Figure 9 & 10)

The sixth period includes the remainder of the mission and will be devoted to the CSM solo operations. The period will include two SPS orbit-shaping burns to lower perigee and raise apogee thereby establishing an orbit which permits SM RCS deorbit should a SPS malfunction occur later in the mission. The remainder of this period preceding deorbit will be devoted to navigation exercises and a multispectral terrain photography scientific experiment. This period and the mission will terminate with an SPS deorbit burn, reentry, and splashdown in the Atlantic recovery area, approximately 1000 nautical miles east of Kennedy Space Center.

APOLLO 9 TENTH DAY

CM/SM SEPARATION

RE-ENTRY

Fig 10

ATLANTIC - SPLASHDOWN

The Apollo 9 crew will be picked up by the Prime Recovery Ship, USS GUADALCANAL, LPH 7 (Landing Platform Helicopter), and will be airlifted by helicopter the following morning to Norfolk, Virginia and subsequently to the Manned Spacecraft Center.

Multispectral Terrain Photography Experiment (SO65)

Photographic experiment SO65 will be conducted during the sixth period of the Apollo 9 mission. The general purpose is to obtain selective multispectral photographs with four different film/filter combinations of selected land and ocean areas. The equipment will consist of four Hasselblad cameras to be carried in the CM.

Photographs acquired during this experiment will be used for scientific analyses in the earth resources disciplines. These photographs are expected to yield information concerning such items as geologic features, water runoff, snow and ice cover, pollution, distribution of soil types, forestry resources, ocean currents, beach erosion, shallow water sediment migration, environmental variations, and weather.

CONTINGENCY OPERATIONS

If an anomaly occurs after lift-off that would prevent the AS-504 space vehicle from following its nominal flight plan, an abort or an alternate mission will be initiated. An abort would provide only for an acceptable CM/crew recovery while an alternate mission would attempt to achieve some of the mission objectives before providing for an acceptable CM/crew recovery.

ABORTS

Launch Aborts

During launch, the velocity, altitude, atmosphere, and launch configuration change rapidly; therefore, several abort modes, each adapted to a portion of the launch trajectory, are required.

Mode I abort procedure is designed for safe recovery of the CM following aborts occurring between Launch Escape System (LES) activation (approximately T minus 30 minutes) and Launch Escape Tower (LET) jettison, approximately 3 minutes GET. The procedure consists of the LET pulling the Command Module (CM) away from the remainder of the space vehicle and propelling it a safe distance down range. The resulting landing point lies between the launch site and approximately 490 nautical miles down range.

The Mode II abort would be performed from the time the LET is jettisoned until the full-lift CM landing point is 3200 nautical miles down range, approximately 10 minutes GET. The procedure consists of separating the CSM combination from the remainder of the space vehicle, separating the CM from the SM, and then letting the CM free fall to entry. The entry would be a full-lift, or maximum range trajectory, with a landing 400 to 3200 nautical miles down range on the ground track.

Mode III abort can be performed from the time the full-lift landing point range reaches 3200 nautical miles until orbital insertion. The procedure would consist of separating the CSM from the remainder of the space vehicle and then, if necessary, performing a retrograde burn with the SPS so that the half-lift landing point is no farther than 3350 nautical miles down range. A half-lift entry would be flown which causes the landing point to be approximately 70 nautical miles south of the nominal ground track between 3000 and 3350 nautical miles down range.

The Mode IV abort procedure is an abort to earth orbit, Contingency Orbit Insertion (COI), and could be performed anytime after the SPS has the capability to insert the CSM into orbit. This capability begins at approximately 10 minutes GET. The procedure would consist of separating the CSM from the remainder of the space vehicle and, two minutes later, performing a posigrade SPS burn to insert the CSM into earth orbit with a perigee of at least 75 nautical miles. The CSM could then remain in earth orbit for an earth orbital alternate mission, or if necessary, return to earth in the West Atlantic or Central Pacific Ocean after one revolution. This mode of abort is preferred over the Mode III abort and would be used unless an immediate return to earth is necessary during the launch phase.

The last abort procedure is an Apogee Kick (AK) Mode. This mode is a variation of the Mode IV wherein the SPS burn to orbit occurs at apogee altitude to raise the perigee to 75 nautical miles. The maneuver is executed whenever the orbital apogee at S-IVB cutoff is favorably situated and the corresponding Mode IV delta V requirement is greater than 100 feet per second. Like the Mode IV contingency orbit insertion, this maneuver is prime when the capability exists, except for those situations where an immediate return to earth is required.

Earth Orbit Aborts

Once the CSM/LM/IU/S-IVB is safely inserted into earth parking orbit, a return to-earth abort would be performed by separating the CSM from the LM/IU/S-IVB and then utilizing the SPS for a retrograde burn to place the CM, after CM/SM separation, on an atmosphere-intersecting trajectory. After entry the crew would fly a guided flight path to a pre-selected target point if possible.

Rendezvous Aborts

A capability will be maintained throughout the rendezvous to provide for non time-critical and time-critical aborts by either the LM or by the CSM.

Alternate Missions

Seven alternate missions have been developed for Apollo 9 which provide for a maximum accomplishment of test objectives while adhering to mission constraints pertaining to mission ground rules, crew safety, and trajectory considerations. A summary of the alternate missions and the precipitating functional failures is shown on Table 3.

Alternate Mission A

Alternate Mission A is a CSM only mission and will be used in the event that a COI is necessary, the LM cannot be ejected from the SLA, or the LM Descent Stage is deemed unsafe. All SPS burns are planned to be accomplished in this alternate; however, duration and scheduling of the burns will be real time decisions.

Alternate Mission B

Alternate Mission B is designed for an SPS failure, CSM lifetime problems, or electrical problems in the LM. Should this alternate be necessary, real time evaluation of the mission will be performed and the crew and events activities will be rescheduled to accomplish a maximum of mission objectives. A SM RCS deorbit is planned into the prime recovery area.

TABLE 3 ALTERNATE MISSION SEQUENCE APOLLO 9 OF EVENTS

NOMINAL MISSION PERIOD OF ENTRY		FUNCTIONAL FAILURE PRECIPITATING ALTERNATE MISSION
		ALTERNATE MISSION A
I	COI	COI
I OR 2	SPS I	LM CANNOT BE EJECTED FROM SLA UNSAFE DESCENT STAGE
	SPS 2	
	SPS 3	
	SPS 4	
3-5	SPS 5	
3-6	SPS 6	
	SPS 7	
	SPS 8	

ALTERNATE MISSION B

I	T, D AND E	SPS FAILURE
2 OR 3	LM SYSTEMS EVALUATION	CSM LIFETIME PROBLEM
3 OR 4	EXECUTE DOCKED DPS BURN PERFORM EVA	DESCENT STAGE ELECTRICAL POWER PROBLEMS
4 OR 5	STATION KEEPING (STAGE LM PRIOR TO DOCKING)	ASCENT STAGE ELECTRICAL POWER PROBLEMS
4 OR 5	LONG APS BURN	
5 OR 6	DEORBIT	

ALTERNATE MISSION C

3 OR 4	PERFORM EVA	LONG APS BURN CONTINUE MISSION UNSAFE DESCENT STAGE EVT TAKES LONGER THAN 15 MINUTES

ALTERNATE MISSION D

I	T, D AND E	CSM LIFETIME PROBLEM
2 OR 3	LM SYSTEMS EVALUATION	
	EXECUTE DOCKED DPS BURN	EITHER CSM COOLANT LOOP FAILS
		DESCENT STAGE ELECTRICAL POWER PROBLEMS
3 - 5	STAGE DESCENT ST49 LONG APS BURN	ASCENT STAGE ELECTRICAL POWER PROBLEMS
3 - 6	DEBRIEF	

ALTERNATE MISSION E

5	E-5A STATION KEEPING CONTINUE NOMINAL MISSION TIMELINE	LM PRIMARY COOLANT LOOP LOST
	E-5B MINI-FOOTBALL RENDEZVOUS CONTINUE NOMINAL MISSION TIMELINE	DESCENT STAGE ELECTRICAL POWER PROBLEM
	E-5C FOOTBALL RENDEZVOUS CONTINUE NOMINAL MISSION TIMELINE	ASCENT STAGE ELECTRICAL POWER PROBLEM
	E-5D CSM ACTIVE RENDEZVOUS CONTINUE NOMINAL MISSION TIMELINE	PGNCS FAILURE
		RENDEZVOUS RADAR FAILURE
		LM AGS LOST
		UNSAFE DESCENT STAGE

ALTERNATE MISSION F

3	DELETE DOCKED BPS BURN PERFORM SPS 5 PERFORM EVA	PGNCS LOST 4
5	STATION KEEP, STAGE LM AND DOCK EXECUTE CSM ACTIVE RENDEZVOUS (E-5D) DELETE LONG DURATION APS BURN CONTINUE MISSION	

ALTERNATE MISSION G

3	DELETE DOCKED BPS BURN PERFORM EVA	DPS NON OPERABLE 4 LM PRIMARY COOLANT LOOP LOST
9	STATION KEEPING (E-5A) LONG APS BURN CONTINUE MISSION	

Alternate Mission C

Alternate Mission C is designed for an unsafe Descent Stage. With an unsafe Descent Stage, mission rules call for jettisoning the Descent Stage if the LM is manned or the entire LM if the LM is unmanned. Without the Descent Stage, the remainder of the LM activities are accomplished on Ascent Stage consumables. Therefore, in order to accomplish a maximum number of mandatory DTO's with the consumables available, undocking the LM is deleted in favor of accomplishing the long APS burn and the EVA. However, if the unsafe Descent Stage is discovered late in the fourth period of activities, sufficient Ascent Stage consumables might exist for the station-keeping and the long APS burn.

If EVT takes longer than 15 minutes, undocked manned LM activities will not be undertaken since a backup for IVT is not available.

Alternate Mission D

Alternate Mission D is designed for failure of a CSM coolant loop, CSM lifetime problems, or LM electrical power problems. This would be a time-critical alternate; therefore, all SPS burns would be deleted in order to accomplish a maximum number of high priority DTO's. Without the SPS burns, the docked DPS burn will be retargeted to provide a deorbit capability into the prime recovery area as soon as possible after the long APS burn. Real time decisions will be necessary to schedule crew activities and additional spacecraft events.

Alternate Mission E

Alternate Mission E is designed for several anomalous situations as listed in Table 3 and consists of a series of modified rendezvous that may be selected to replace the nominal rendezvous in activity period five. The

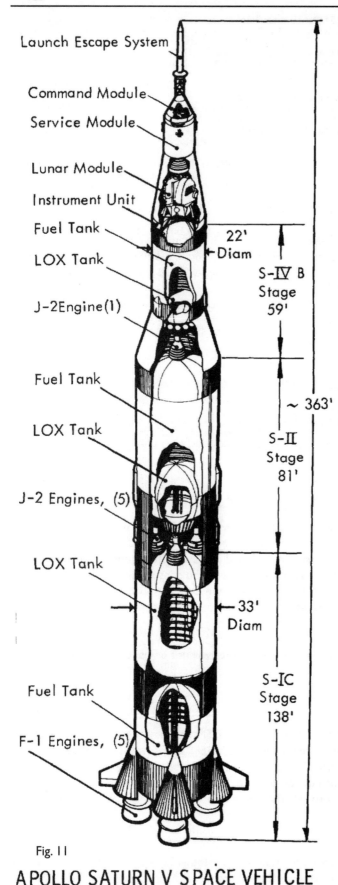

Fig. 11

APOLLO SATURN V SPACE VEHICLE

Launch Escape System

Command Module

Service Module

Lunar Module

Instrument Unit

Fuel Tank

LOX Tank

J-2 Engine (1)

Fuel Tank

LOX Tank

J-2 Engines, (5)

LOX Tank

Fuel Tank

F-1 Engines, (5)

22' Diam

S-IV B Stage 59'

~ 363'

S-II Stage 81'

33' Diam

S-IC Stage 138'

events prior to and after the modified rendezvous are nominal. The selection of the modified rendezvous is made in real time and will depend on the failure precipitating the alternate mission and the resulting consumables status. The four modified rendezvous plans are:

E-5a.   Station-keeping
E-5b.   Mini-football rendezvous
E-5c.   Football rendezvous
E-5d.   CSM-active rendezvous

With the exception of the CSM-active rendezvous, the modified rendezvous are portions of the nominal rendezvous found in the Detailed Flight Mission Description. The LM is staged prior to docking with the CSM in the LM-active modified rendezvous in order to satisfy the test conditions of the LM-active docking DTO.

Alternate Mission F

Alternate Mission F is designed for a LM PGNCS failure. A slightly modified sequence of events results, with the docked DPS burn, LM-active rendezvous, and long APS burn deleted. A CSM-active rendezvous is added. The LM-active rendezvous is not attempted since a backup guidance system is not available. The docked DPS burn is deleted since the AGS lacks moment control of the Ascent Stage/CSM stacked configuration. The long APS burn is deleted as a result of the AGS lacking ground-commanded shutdown capability. The SPS-5 burn is executed to provide a circular orbit for the CSM-active rendezvous.

The unmanned LM is left as a target for the CSM in the CSM-active rendezvous. However, without a man in the secondary suit loop, the suit loop water boiler will freeze because of the lack of heat input to the boiler. This will eventually cause the AGS to fail . Hence, it is not known if the LM lights will be visible (LM might be tumbling) at TPI. A real time decision will have to be made at that point to continue or delete the terminal phase maneuvers.

Both the SPS-5 burn with a heavy LM and a CSM-active rendezvous result in a greater propellant usage than nominal. However, both SM RCS and SPS propellants should still be within their current redlines. A real time consumables analysis will be run if there is any off-nominal SPS or SM RCS performance prior to entry into the alternate mission.

The LM is staged just prior to docking with the CSM following the station-keeping exercise. This satisfies the test conditions of the LM-active docking DTO.

Alternate Mission G

Alternate Mission G is designed to accommodate a non-operable DPS or a LM primary coolant loop failure. A non-simultaneous eat-rest-eat cycle is included in the mission since there are specific systems failures that lead to the general functional failure that require continuous monitoring by at least one crewman. If the specific failure does not require this monitoring, the events can be rescheduled in real time.

The sequence of events is slightly modified with the docked DPS burn and the LM-active rendezvous deleted. The docked DPS burn is deleted in case of the LM coolant loop failure since the PGNCS would be uncooled during a manned burn. The long APS burn, however, is accomplished since the LM is unmanned and the ground has shutdown capability in case the PGNCS fails from overheating and the LM starts to tumble. Since this is an alternate mission designed for a LM failure, the nominal sequence of events is picked up following the long APS burn. The LM is staged just prior to docking with the CSM in the general Alternate Mission G.

SPACE VEHICLE DESCRIPTION

The Apollo 9 Mission will be performed by an Apollo Saturn V Space Vehicle (Figure 11) designated AS-504, which consists of a three-stage Saturn V Launch Vehicle, and a complete Apollo Block II Spacecraft. A more comprehensive description of the space vehicle and its subsystems is included in the Mission Operation Report Supplement. The following is a brief description of the various stages of Apollo 9.

The Saturn V Launch Vehicle (SA-504) consists of three propulsion stages (S-IC, S-II, S-IVB) and an Instrument Unit (IU). The Apollo Spacecraft payload for Apollo 9 consists of a Launch Escape System (LES), Block II Command and Service Module (CSM 104), a Spacecraft LM Adapter (SLA 12), and a Lunar Module (LM-3). A list of current weights for the space vehicle is contained in Table 4.

TABLE 4 APOLLO 9 WEIGHT SUMMARY (All Weights in Pounds)

STAGE/ MODULE	INERT WEIGHT	TOTAL EXPENDABLES	TOTAL WEIGHT	FINAL SEPARATION WEIGHT
S-IC Stage	295,200	4,736,330	5,031,530	369,640
S-IC/S-II Interstage	11,665		11,665	
S-II Stage	84,600	979,030	1,064,630	96,540
S-II/IVB Interstage	8,080		8,080	
S-IVB Stage	25,300	233,860	259,160	28,700
Instrument Unit	4,270		4,270	
Launch Vehicle at Ignition	-	-	6,379,335	
SCAM Adapter	4,105		4,105	
Lunar Module	10,165	21,860	32,025	
Service Module	11,295	35,305	46,600	13,075
Command Module	12,405		12,405	11,135
Launch Escape System (Splashdown)	8,850		8,850	
Spacecraft at Ignition	-	-	103,985	
Space Vehicle at Ignition			6,483,320	
S-IC Thrust Buildup			- 86,265	
Space Vehicle at Lift-off			6,397,055	
Space Vehicle at Orbit Insertion			289,970	

LAUNCH VEHICLE DESCRIPTION

First Stage (S-IC)

The S-IC is powered by five F-1 rocket engines each developing approximately 1,522,000 pounds of thrust at sea level and building up to 1.7 million pounds before cutoff. One engine, mounted on the vehicle longitudinal centerline, is fixed; the remaining four engines, mounted in a square pattern about the center line, are gimbaled for thrust vector control by signals from the control system housed in the IU. The F-1 engines utilize LOX (liquid oxygen) and RP-1 (kerosene) as propellants.

Second Stage (S-II)

The S-II is powered by five high-performance J-2 rocket engines each developing approximately 230,000 pounds of thrust in a vacuum. One engine, mounted on the vehicle longitudinal centerline, is fixed; the remaining four engines, mounted in a square pattern about the centerline, are gimbaled for thrust vector control by signals from the control system housed in the IU. The J-2 engines utilize LOX and LH2 (liquid hydrogen) as propellants.

The SA-504 is the first Saturn V to utilize the light weight S-II stage. This stage is approximately 4000 pounds lighter than the S-II stage flown on Apollo 8. This reduction is the result of a concerted effort to reduce the overall weight of the Saturn V Launch Vehicle.

Third Stage (S-IVB)

The S-IVB is powered by a single J-2 engine developing approximately 230,000 pounds of thrust in a vacuum. As installed in the S-IVB, the J-2 engine features a multiple start capability. The engine is gimbaled for thrust vector control in pitch and yaw. Roll control is provided by the Auxiliary Propulsion System (APS) modules containing motors to provide roll control during main stage operations and pitch, yaw, and roll control during non-propulsive orbital flight.

Instrument Unit

The Instrument Unit (IU) contains the following: Electrical system, self-contained and battery powered; Environmental Control System, provides thermal conditioning for the electrical components and guidance systems contained in the assembly; Guidance and Control System, used in solving guidance equations and controlling the attitude of the vehicle; Measuring and Telemetry System, monitors and transmits flight parameters and vehicle operation information to ground stations; Radio Frequency System, provides for tracking and command signals; components of the Emergency Detection System (EDS).

SPACECRAFT DESCRIPTION

Command Module

The Command Module (CM) (Figure 12) serves as the command, control, and communications center for most of the mission. Supplemented by the SM, it provides all life support elements for three crewmen in the mission environments and for their safe return to earth's surface. It is capable of attitude control about three axes and some lateral lift translation at high velocities in earth atmosphere. It also permits LM attachment, CM/LM ingress and egress, and serves as a buoyant vessel in open ocean.

Service Module

The Service Module (SM) (Figure 13) provides the main spacecraft propulsion and maneuvering capability during the mission. The Service Propulsion System (SPS) provides up to 20,500 pounds of thrust in a vacuum. The Service Module Reaction Control System (SM RCS) provides for maneuvering about and along three axes. The SM provides most of the spacecraft consumables (oxygen, water, propellant, hydrogen). It

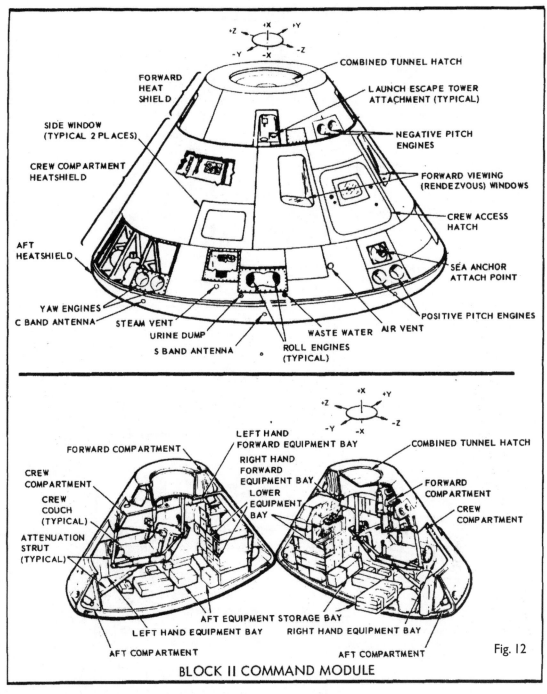

FORWARD HEAT SHIELD

COMBINED TUNNEL HATCH

LAUNCH ESCAPE TOWER ATTACHMENT (TYPICAL)

SIDE WINDOW (TYPICAL 2 PLACES)

NEGATIVE PITCH ENGINES

CREW COMPARTMENT HEATSHIELD

FORWARD VIEWING (RENDEZVOUS) WINDOWS

CREW ACCESS HATCH

AFT HEATSHIELD

SEA ANCHOR ATTACH POINT

YAW ENGINES

C BAND ANTENNA

STEAM VENT

URINE DUMP

S BAND ANTENNA

ROLL ENGINES (TYPICAL)

WASTE WATER

AIR VENT

POSITIVE PITCH ENGINES

FORWARD COMPARTMENT

LEFT HAND FORWARD EQUIPMENT BAY

RIGHT HAND FORWARD EQUIPMENT BAY

LOWER EQUIPMENT BAY

COMBINED TUNNEL HATCH

CREW COMPARTMENT

CREW COUCH (TYPICAL)

ATTENUATION STRUT (TYPICAL)

FORWARD COMPARTMENT

CREW COMPARTMENT

AFT EQUIPMENT STORAGE BAY

LEFT HAND EQUIPMENT BAY

RIGHT HAND EQUIPMENT BAY

AFT COMPARTMENT

AFT COMPARTMENT

Fig. 12

BLOCK II COMMAND MODULE

supplements environmental, electrical power, and propulsion requirements of the CM. The SM remains attached to the CM until it is jettisoned just before CM re-entry.

Common Command and Service Module Systems

There are a number of systems which are common to the CM and SM.

Guidance and Navigation System

The Guidance and Navigation (G&N) System measures spacecraft attitude and acceleration, determines trajectory, controls spacecraft attitude, controls the thrust vector of the SPS engine, and provides abort information and display data.

SERVICE MODULE

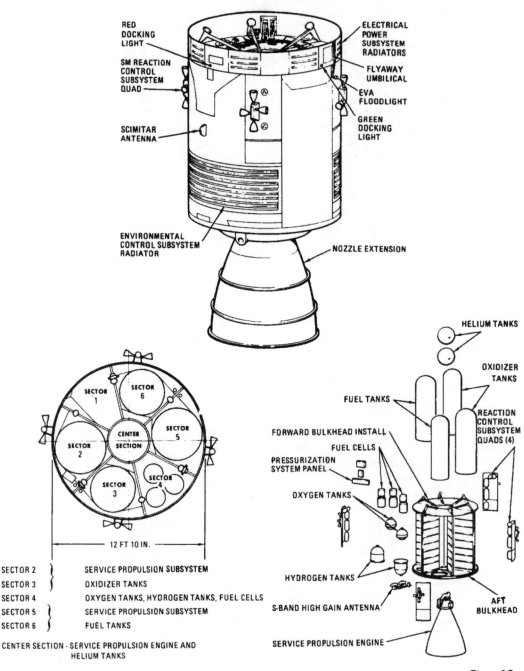

RED DOCKING LIGHT

SM REACTION CONTROL SUBSYSTEM QUAD

SCIMITAR ANTENNA

ELECTRICAL POWER SUBSYSTEM RADIATORS

FLYAWAY UMBILICAL

EVA FLOODLIGHT

GREEN DOCKING LIGHT

ENVIRONMENTAL CONTROL SUBSYSTEM RADIATOR

NOZZLE EXTENSION

SECTOR 1
SECTOR 6
SECTOR 5
SECTOR 2
CENTER SECTION
SECTOR 3
SECTOR 4

12 FT 10 IN.

SECTOR 2	}	SERVICE PROPULSION SUBSYSTEM
SECTOR 3		OXIDIZER TANKS
SECTOR 4		OXYGEN TANKS, HYDROGEN TANKS, FUEL CELLS
SECTOR 5	}	SERVICE PROPULSION SUBSYSTEM
SECTOR 6	}	FUEL TANKS

CENTER SECTION - SERVICE PROPULSION ENGINE AND HELIUM TANKS

HELIUM TANKS

OXIDIZER TANKS

FUEL TANKS

REACTION CONTROL SUBSYSTEM QUADS (4)

FORWARD BULKHEAD INSTALL

FUEL CELLS

PRESSURIZATION SYSTEM PANEL

OXYGEN TANKS

HYDROGEN TANKS

S-BAND HIGH GAIN ANTENNA

AFT BULKHEAD

SERVICE PROPULSION ENGINE

Fig. 13

Stabilization and Control System

The Stabilization and Control System (SCS) provides control and monitoring of the spacecraft attitude, backup control of the thrust vector of the SPS engine and a backup inertial reference.

Reaction Control System

The Reaction Control System (RCS) provides thrust for attitude and small translational maneuvers of the spacecraft in response to automatic control signals from the SCS in conjunction with the G&N system. The

CM and SM each has its own independent and redundant system, the CM RCS and the SM RCS respectively. Propellants for the RCS are hypergolic.

Electrical Power System

The Electrical Power System (EPS) supplies all electrical power required by the CSM. The primary power source is located in the SM and consists of three fuel cells which are the prime spacecraft power from lift-off through CM/SM separation. Five batteries -- three for peak load intervals, entry and post-landing, and two for pyrotechnic uses -- are located in the CM.

Environmental Control System

The Environmental Control System (ECS) provides a controlled cabin environment and dispersion of CM equipment heat loads.

Telecommunications System

The Telecommunications (T/C) System provides for the acquisition, processing, storage, transmission and reception of telemetry, tracking, and ranging data among the spacecraft and ground stations.

Sequential System

Major Sequential Subsystems (SEQ) are the Sequential Events Control System (SECS), Emergency Detection System (EDS), Launch Escape System (LES), and Earth Landing System (ELS). The subsystems interface with the RCS or SPS during an abort.

Spacecraft LM Adapter

The Spacecraft LM Adapter (SLA) is a conical structure which provides a structural load path between the LV and SM and also supports the LM. Aerodynamically, the SLA smoothly encloses the SM engine nozzle and irregularly-shaped LM, and transitions the SV diameter from that of the upper LV stage to that of the SM. The upper section is made up of four panels that swing open at the top and are jettisoned away from the spacecraft by springs attached to the lower fixed panels.

Lunar Module

Lunar Module (LM) 3 for the Apollo 9 Mission will exercise, in earth orbit, many of the systems and capabilities of its prime mission as a lunar landing and launching vehicle. The LM (Figure 14) is a two-stage vehicle designed to transport two crewmen from a docked position with the CSM to the lunar surface, serve as a base for lunar surface crew operations, and to provide for their safe return to the docked position. The upper stage is termed the Ascent Stage (AS) and the lower stage, the Descent Stage (DS). In the nominal mission, the two stages are operated as a single unit until the lunar landing is accomplished. The Ascent Stage is used for ascent from the lunar surface and rendezvous with the CSM. The LM's main propulsion includes a gimbaled, throttleable Descent Propulsion System (DPS) engine and a fixed, non-throttleable Ascent Propulsion System (APS) engine. A 16-jet Reaction Control System on the Ascent Stage provides for stabilization and maneuvering. All propulsive systems utilize storable hypergolic propellants. The Guidance, Navigation, and Control System (GN&CS) has the capability to automatically implement all parameters required for safe landing from lunar orbit and accomplish a CSM-LM rendezvous from lunar launch. Landing and Rendezvous radar systems aid the GN&CS system. The Instrument System provides for LM systems checkout and displays data for monitoring or manually controlling LM systems. The environmental Control System provides a satisfactory environment for equipment and human life. The Electrical Power System relies upon four batteries in the Descent Stage and two batteries in the Ascent Stage when undocked from the CSM. Electrical Power is provided by the CSM when the LM is docked. Telecommunications is provided to the MSFN, the CSM, and an extravehicular astronaut.

LUNAR MODULE

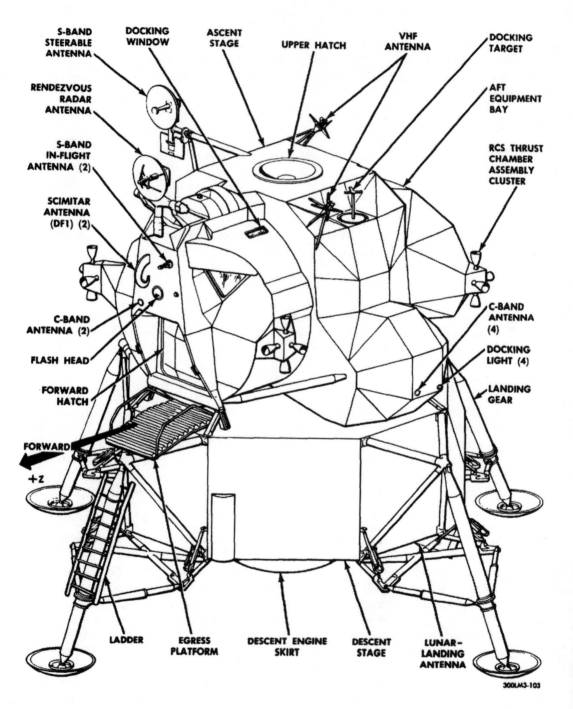

S-BAND STEERABLE ANTENNA
DOCKING WINDOW
ASCENT STAGE
UPPER HATCH
VHF ANTENNA
DOCKING TARGET
RENDEZVOUS RADAR ANTENNA
AFT EQUIPMENT BAY
S-BAND IN-FLIGHT ANTENNA (2)
RCS THRUST CHAMBER ASSEMBLY CLUSTER
SCIMITAR ANTENNA (DFI) (2)
C-BAND ANTENNA (2)
FLASH HEAD
C-BAND ANTENNA (4)
DOCKING LIGHT (4)
LANDING GEAR
FORWARD HATCH
FORWARD
+Z
LADDER
EGRESS PLATFORM
DESCENT ENGINE SKIRT
DESCENT STAGE
LUNAR-LANDING ANTENNA
300LM3-103

Fig. 14

The LM can operate for 48 hours after separation from the CSM. Insulation provides protection against 350°F temperatures and in certain required areas up to 1000°F. A detailed description of the LM and its systems is in the MOR Supplement.

Television Camera

Apollo 9 will provide the first operational test of the television camera designed for eventual use by the lunar

1966 crew shot. David Scott, Jim McDivitt and Russel Schweickart (Above).

Apollo 5 LM during final assembly, the only LM to fly before Apollo 9 (Left).

Preflight simulator training (Above right).

Apollo 9 LM during final assembly (Left).

Apollo 9 Saturn V during assembly in the VAB.

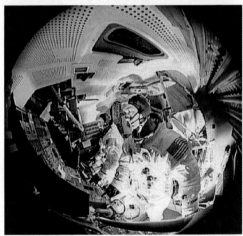

The prime flight crew with their vehicle in the
background. (Above right)
The Saturn V Booster on the crawlerway (Above left)
Apollo 9 on it's way to Earth orbit (left and below)

The Lunar Module nestled
inside the Saturn IVB after the
panel ejection. (Left)
Shortly after this picture was
taken the Command Module
docked with the Lunar Module
and pulled it clear of the S-IVB.

The Lunar Module inside the S-IVB just prior to docking.

A composed McDivitt during the flight, Schweickart donning his suit and Scott checking out the view.

Scott leaning out of the CM to retrieve scientific
samples, the LM in the foreground. (Above)
"Gumdrop" during the separation and docking
maneuvers. (Below and left)

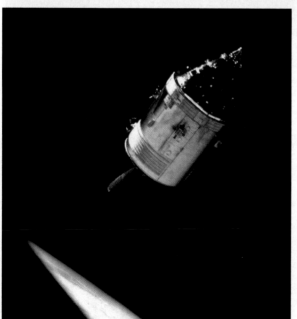

The Command
Module "Gumdrop"
(Above) and it's
pirouetting partner
the Lunar Module
"Spider"
(Left and right)

Humanities first true space vehicle, the Lunar Module "Spider", is run through it's paces by Commander McDivitt and Pilot Schweickart.

Another shot of "Spider" against the clouds below, clearly showing the long needle-like sensors extending from each foot-pad. On future missions to the moon these sensors made contact with the moon's surface and activated a contact light in the cockpit — the pilot would then shut down the descent engine and let the vehicle fall the last few feet to the lunar surface.

The LM returning without the descent stage. This inverted shot clearly shows the ascent engine which would later be used to lift-off from the moon.

Astronaut Schweickart performing the first space-test of the suit which would later be used to walk on the moon.
Note the reflection of the earth in his visor.

Several more shots showing Schweickart during his EVA (space-walk). Note the LM footpad below him and the long white metal hand-rail designed to accomodate hand-over-hand movement to the Command Module (left).

Home again, Schweickart and McDivitt salute the crew of the USS Guadalcanal (above)

The crew with a huge cake aboard USS Guadalcanal. (left)

landing crewmen. This is a new design camera that has not been flown on any previous mission. The camera will be carried in the ascent stage of the LM and will not be operated from inside the CM. The first of the two scheduled transmissions will be of the LM interior. In the second transmission, the camera will be operated by the LMP at the end of his extravehicular activity period and will feature pictures of the CSM/LM exterior.

Launch Escape System

The Launch Escape System (LES) provides the means for separating the CM from the LV during pad or sub-orbital aborts through completion of second stage burn. This system consists primarily of the Launch Escape Tower (LET), Launch Escape Motor, Tower Jettison Motor and Pitch Motor. All motors utilize solid propellants. A Boost Protective Cover (BPC) is attached to the LET and covers the CM from LES rocket exhaust and also from aerodynamic heat generated during LV boost.

CONFIGURATION DIFFERENCES

The space vehicle for Apollo 9 varies in its configuration from that flown on Apollo 8 and those to be flown on subsequent missions. These differences are the result of the normal growth, planned changes, and experience gained on previous missions. Following is a listing of the major configuration differences between AS-503 and AS-504. LM-3 is compared with LM-I, which was flown on Apollo 5.

S-IC STAGE

Deleted film camera system
Reduced R&D instrumentation
Installed redesigned F-I engine injector
Removed television cameras
Increased propulsion performance
Reduced weight by removal of forward skirt insulation and revising "Y" rings and skin taper in propellant tanks

S-II STAGE
First flight of lightweight structure
Redesigned separation planes tension plates
Uprated J-2 engines
Reinforced thrust structure
Changed PU system to closed loop

S-IVB STAGE
Reduced Instrumentation battery capacity
Deleted anti-flutter kit
Uprated J-2 engine

INSTRUMENT UNIT
Enlarged methanol accumulator
Changed networks to disable spacecraft control of launch vehicle
Removed one instrument battery
Deleted S-band telemetry

COMMAND MODULE
Added forward hatch emergency closing link
Added general purpose timer
Added precured RTV to side and hatch windows
Added SO65 camera experiment equipment
Added docking probe, ring, and latches
Added RCS propulsion burst disc
Added Solenoid valve to RCS propellant system
Changed S-band power amplifier configuration to 0006 configuration
Deleted flight qualification recorder

LUNAR MODULE
First operational flight of Oxygen Supply Module
First operational flight of Water Control Module
First flight of VHF transceiver and Diplexer
First flight to use exterior tracking light
First flight to use ascent engine arming assembly
First operational flight of the Abort Guidance Section
First operational flight of the Rendezvous Radar
First flight of the landing radar electronic and antenna assembly
First flight using thrust translation controller assembly
First flight to use Orbital Rate Drive
Modified CO_2 partial pressure sensor to correct EMI, vibration, and outgassing problems
Added high-reliability transformer for use with the S-band steerable antenna
Added pressure switch to RCS
Modified thermal insulation in the Rendezvous Radar Antenna Assembly
Installed Landing gear
Added high-efficiency reflective coated cabin and docking windows
Added split AC bus
Added more reliable Signal Processor Assembly
Added manual trim shutdown to descent engine control assembly
Modified Stabilization and Control Assembly No. I to eliminate single failure point
Added fire preventive and resistive materials
Added TV camera

SPACECRAFT LM ADAPTER
Redesigned SLA panel charges
Added spring ejector for LM
Added LM separation sequence controllers
Deleted POGO instrumentation
Added "cookie cutter" emergency egress equipment (was included on Apollo 5)

HUMAN SYSTEM PROVISIONS

The major human system provisions included for the Apollo 9 mission are: Space Suits, Bio-instrumentation System, Medical Provisions, Crew Personal Hygiene, Crew Meals, Sleeping Accommodations, Oxygen Masks, and Survival Equipment. These systems provisions are described in detail in the Mission Operation Report Supplement.

LAUNCH COMPLEX

The AS-504 Space Vehicle (SV) will be launched from Launch Complex (LC) 39 at the Kennedy Space Center. The major components of LC 39 include the Vehicle Assembly Building (VAB), the Launch Control Center (LCC), the Mobile Launcher (ML), the Crawler Transporter (C/T), the Mobile Service Structure (MSS), and the Launch Pad.

The LCC is a permanent structure located adjacent to the VAB and serves as the focal point for monitoring and controlling vehicle checkout and launch activities for all Saturn V launches. The ground floor of the structure is devoted to service and support functions. Telemetry equipment occupies the second floor and the third floor is divided into firing rooms, computer rooms, and offices. Firing room 2 will be used for Apollo 9.

The AS-504 SV was received at KSC and assembly and initial overall checkout was performed in the VAB on the mobile launcher. Rollout occurred on 3 January 1969. Transportation to the pad of the assembled SV and ML was provided by the Crawler Transporter (C/T) which also moved the MSS to the pad after the ML and SV had been secured. The MSS provides 360-degree access to the SV at the launch pad by means of five vertically-adjustable, elevator-serviced, enclosed platforms. The MSS will be removed to its park position prior to launch.

The emergency egress route system at LC 39 is made up of three major components: the high speed elevators, slide tube, and slide wire. The primary route for egress from the CM is via the elevators and, if necessary, through the slide tube which exits into an underground blast room.

A more complete description of LC 39 is in the MOR Supplement.

MISSION SUPPORT

Mission support is provided by the Launch Control Center (LCC), the Mission Control Center (MCC), the Manned Space Flight Network (MSFN), and the recovery forces. The LCC is essentially concerned with pre-launch checkout, countdown, and with launching the SV, while MCC located at Houston, Texas, provides centralized mission control from lift-off through recovery. MCC functions within the framework of a Communications, Command, and Telemetry System (CCATS); Real Time Computer Complex (RTCC); Voice Communications System; Display/control System; and, a Mission Operations Control Room (MOCR). These systems allow the flight control personnel to remain in contact with the spacecraft, receive telemetry and operational data which can be processed by the CCATS and RTCC for verification of a safe mission or compute alternatives. The MOCR is staffed with specialists in all aspects of the mission who provide the Mission Director and Flight Director with real time evaluations of mission progress.

The MSFN is a worldwide communications network which is controlled by the MCC during Apollo missions. The network is composed of fixed stations (Figure 15) and is supplemented by mobile stations (Table 5) which are optimally located within a global band extending from approximately 40° South latitude to 40° North latitude. Station capabilities are summarized in Table 6.

The functions of these stations are to provide tracking, telemetry, command and communications both on an updata link to the spacecraft and on a down data link to the MCC. Connection between these many MSFN stations and the MCC is provided by NASA Communications Network (NASCOM). More detail on Mission Support is in the MOR Supplement.

TABLE 5   MSFN MOBILE FACILITIES

APOLLO SHIPS (4 required)

FUNCTION	SUPPORT	LOCATION	NAME
Apollo Insertion Ship	Insertion, abort contingencies	32° N, 45°W	USNS VANGUARD
Apollo Injection Ship	Coverage for CDH and SPS-8	22°N, 131°W	USNS REDSTONE
Apollo Injection Ship	Coverage for phasing	22°S, 160°W	USNS MERCURY
Apollo Reentry Ship	Coverage for TPF	7°S, 170°E	USNS HUNTSVILLE

APOLLO AIRCRAFT (5 required)

ARIA will support the mission on specified revolutions from assigned Test Support Positions (TSP). In addition, ARIA will cover reentry (400,000 ft) through crew recovery. ARIA #1, #2, and #3 will operate in the Pacific Ocean and ARIA #4 and #5 in the Atlantic Ocean.

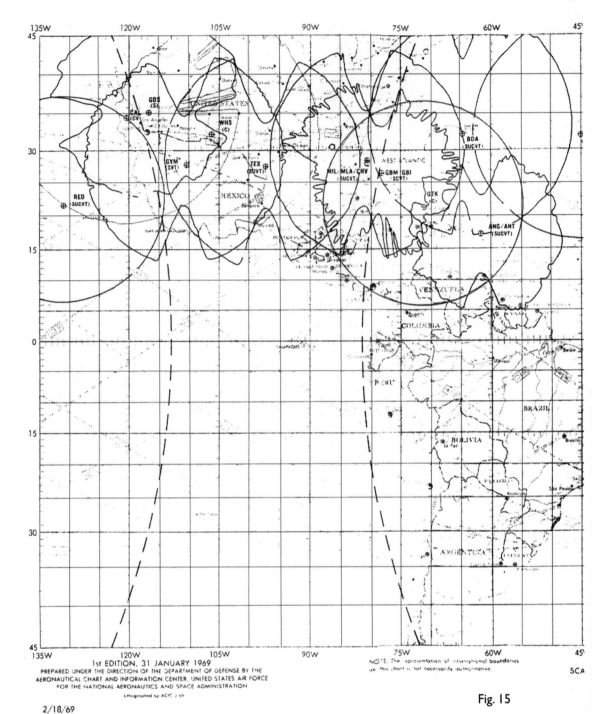

1st EDITION, 31 JANUARY 1969
PREPARED UNDER THE DIRECTION OF THE DEPARTMENT OF DEFENSE BY THE
AERONAUTICAL CHART AND INFORMATION CENTER, UNITED STATES AIR FORCE
FOR THE NATIONAL AERONAUTICS AND SPACE ADMINISTRATION
Lithographed by ACIC 7 69

Fig. 15

2/18/69

RECOVERY SUPPORT PLAN

GENERAL

The primary responsibilities of the recovery forces in supporting the mission are:

A. Rapid location and safe retrieval of the flight crew and spacecraft.

B. The collection, preservation, and return of test data, test hardware, and information relating to the recovery operation.

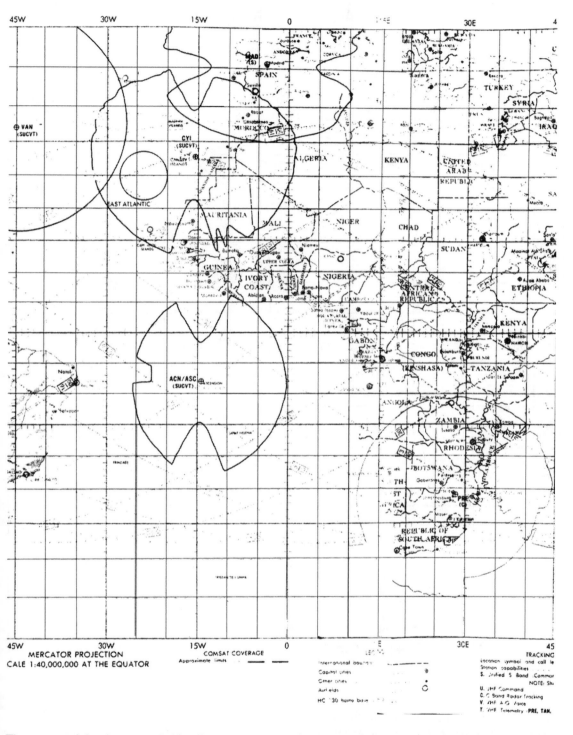

This responsibility begins with lift-off and ends with safe return of the spacecraft and the flight crew to a designated point within the continental United States. The SM will probably break up during reentry and the landing areas will be chosen with this in mind to prevent land impact of parts of the SM.

The recovery planning has been based upon an approximately 10-day duration mission with recovery forces deployed in four recovery zones (1, 2, 3, and 4) as shown in Figure 16 with the primary landing area located in Zone 1. Table 7 provides a summary of recovery forces.

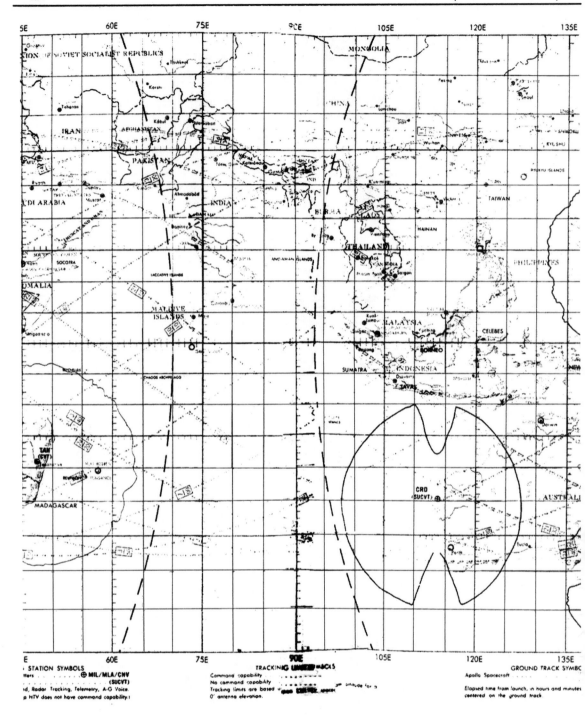

The recovery will be directed from the Recovery Control Room of the MCC, and will be supported by two satellite recovery control centers: The Atlantic Recovery Center located at Norfolk, Virginia, and the Pacific Control Center located at Kunio in the Hawaiian Islands. In addition to the recovery control centers, there will be NASA representatives deployed with recovery forces throughout the worldwide DOD recovery network, at vital staging bases, and in the landing areas to give on-scene technical support to the DOD forces.

RECOVERY GUIDELINES

Recovery guidelines are based upon lighting conditions and the weather in the recovery area. It is highly desirable to have as many daylight hours after landing as is possible in the planned landing area. In addition,

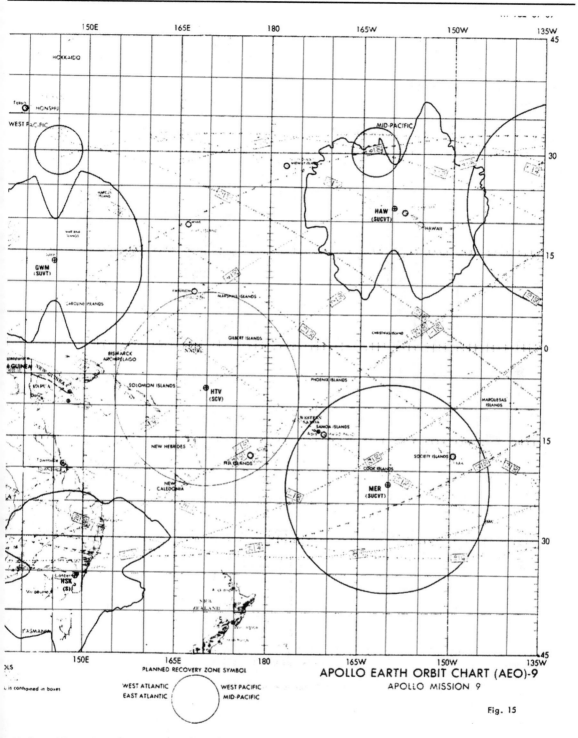

APOLLO EARTH ORBIT CHART (AEO)-9
APOLLO MISSION 9

Fig. 15

it is desirable to have at least two hours of daylight remaining following a landing at the maximum extent of the launch abort area.

The weather guidelines for the launch site, primary, and secondary landing areas are as follows:

A. Surface winds - 25 knots maximum
B. Ceiling - 1500 feet minimum
C. Visibility - 3 nautical miles minimum
D. Wave Height - 8 feet maximum

TABLE 6

Network Configuration for the AS-504 Mission

Systems / Facilities	TRACKING					USB				TLM						CMD			DATA PROCESSING			COMM				OTHER		REMARKS
	C-band (High Speed)	C-band (Low Speed)	ODOP	USB	Optical	TV to MCC	Voice	TLM	Command	VHF Links	FM Recording	ACME Tape Recording	Decoms	Decoms	UHF Commanding	Cmd Destruct	Cmd ship	HSR TLM	ML/CCD	CDP	High-Speed Data	TTY	Voice SCAMA	Voice VHF A/G	Recovery	SPAN	
PAT CNV	2 1	2 1	1		1										3 1	1											Note 1 / Note 1
MLA MIL	2	2		2		3	2	2	2	2	2	2	2			2		2	2		2	2	2	2			Note 1
GBI GRM	2	2		2		2	2	2	2	1 2	1 2	2	2		3 3	1	3	2	2		2	2	2	2			Note 1 / Note 2
ANT ANG	3	3		2		2	2	2	2	2	2	2	2		3	2		2 2	2		2 2	2 2	2 2	2			
BDA CYI	2	2 3		2 3		3	2 3	2 3	2 3	2 3	2 3	2 3	2 3	2	2 3	1		3 3	3		3	3 3	3 3	3	1	3	Note 1
ASC ACN		3		3			3	3	3	3	3	3	3		3			3	3		3	3	3	3			
PRE TAN		3 3					3	3			3										3	3 3	3 3	3			
CRO GWM	3	3 3		3 3			3 3	3 3	3 3	3 3	3 3	3 3	3 4		3 3			3 3	3 3		3	3 3	3 3	3 3	3		
HAW CAL	3	3 3		3			3 3	3 3	3	3	3 3	3 3			3			3 3			3	3 3	3 3	3 3			
GDS GYM		3 3		3 3		3	3 3	3 3	3 3	3	3 3	3 3	3 3					3 3	3 3		3	3 3	3 3	3			
WHS TEX		3		3			3 3	3 3	3 3	3	3 3	3 3			3			3 3	3		3	3 3	3 3	3			
LIMA HSK		3		3			3 3	3 3	3	3 3	3 3							3 3	3		3	3 3	3 3		3		
MAD ARIA (8)		3 3		3 3		3	3 3	3 3	3	3 3	3 3							3 3			3	3 3	3 3				
RED VAN	2	3 2		3 2			3 2	3 2	3 2	3 2	3 2	3 2	3 2	2	3 2			3 2	3 2	3 2	3 2	3 2	3 2	3 2			Note 3 / Note 3
MER HTV		3 3		3 3			3 3	3 3	3 3	3 3	3	3 3	3 3		3			3 3	3 3	3	3	3 3	3 3	3 3			Note 4 / Note 5

Legend: 1. Launch
2. Launch & Orbit
3. Orbit

Note:
1. Launch Abort Contingency   4. Alternate Re-entry/Injection
2. Launch and 4 Revs   5. Re-entry Ship
3. Insertion Ship

TABLE 7 RECOVERY FORCES, APOLLO 9

RECOVERY DESIGNATION	TYPE OF SHIP AND HULL NO.	NAME OF SHIP	HOME PORT
		ATLANTIC OCEAN SHIPS	
PRS	LPH-7	GUADALCANAL	NORFOLK
SRS-1	AIS	VANGUARD	PORT CANAVERAL
SRS-2	LKA-54	ALGOL	NORFOLK

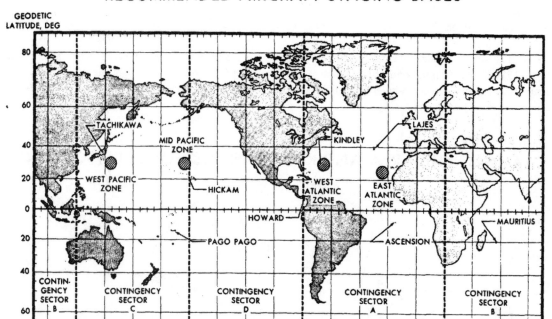

NASA-S-68-6476
Fig. 16

RECOVERY ZONES, CONTINGENCY AREA AND RECOMMENDED AIRCRAFT STAGING BASES

PACIFIC OCEAN SHIPS

SRS-3	DD-852	MASON	YOKOSUKA
SRS-4	DD-449	NICHOLAS	PEARL
SRS-5	DDG-21	COCHRANE	PEARL

18 HC-130 AIRCRAFT

1. KINDLEY AFB, BERMUDA
2. LAJES AFB, AZORES
3. ASCENSION ISLAND
4. MAURITIUS ISLAND
5. PERTH, AUSTRALIA
6. TACHIKAWA AB, JAPAN
7. PAGO PAGO, SAMOA
8. HICKAM AFB, HAWAII
9. HOWARD AFB, PANAMA

RECOVERY AREAS

To define levels of recovery support, spacecraft landing areas are discussed in five general categories: launch site, launch abort, primary, secondary, and contingency. Primary recovery ship support coverage will be required within the primary landing area.

Launch Site Landing Area

A landing could occur in the launch site landing area (Figure 17) if an abort occurred between LES activation and T+90 seconds which corresponds to approximately 41 nautical miles downrange. T+90 seconds is the interface between the launch site recovery forces and deep-water recovery forces. Launch site recovery forces, to the limit of their capability, will support the deep-water recovery forces if such assistance is required. The possible CM landing points lie in a corridor within the launch site area. This corridor, which is determined by the wind profile, will be defined at launch time.

Launch Abort Landing Area

The launch abort landing area is the area in which the CM will land following an abort initiated during the launch phase of flight (Figure 18). The launch abort landing area is a continuous area 50 nautical miles to either side of the ground track extending from the end of the launch site area to 3200 nautical miles downrange. It also includes an area centered approximately 60 nautical miles south of the ground track at 3200 nautical miles downrange which encompasses the Mode III abort landing points. The required access time for aircraft for the launch abort areas is four hours. The retrieval time for ships in the launch abort areas is 30 hours.

Primary Landing Area

The normal end-of-mission area is the primary landing area and requires primary recovery ship support (Figure 19). All aircraft in the end-of-mission area are required to be on station 15 minutes prior to spacecraft reentry to provide direction-finding capability. Access time should not exceed two hours. The Atlantic recovery area will be prime for Apollo 9.

NASA-S-68-6463

LAUNCH SITE AREA AND RECOMMENDED FORCE DEPLOYMENT

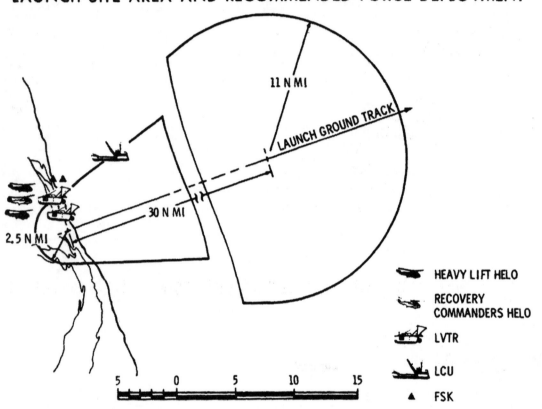

Fig. 17

Secondary Landing Area

A secondary landing area is an area in which a landing could occur after completion of the launch phase when the primary landing area cannot be reached. The probability of a landing in this area is sufficiently high to warrant a requirement for at least secondary recovery ship support. The majority of the secondary landing areas is located in the following recovery zones:

ZONE	ZONE CENTER COORDINATES		RADIUS (N.Mi.)
West Atlantic	28°N	60°W	240
East Atlantic	28°N	25°W	240
West Pacific	28°N	140°E	240
Mid-Pacific	28°N	155°E	240

LAUNCH ABORT AREA AND
RECOMMENDED FORCE DEPLOYMENT

Fig. 18

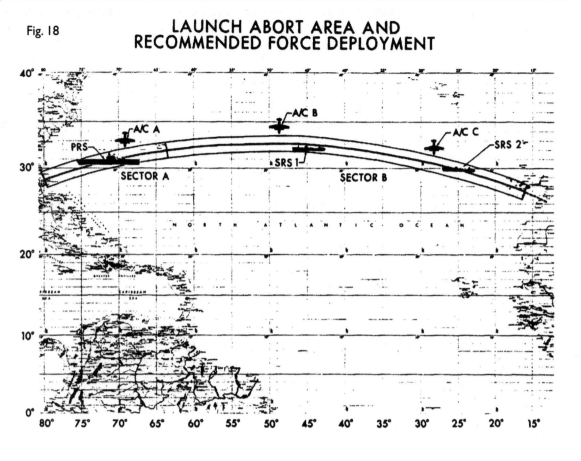

NASA-S-68-6467

RECOMMENDED PRIMARY RECOVERY FORCE DEPLOYMENT

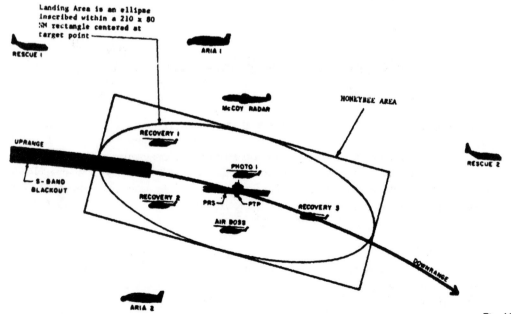

Fig. 19

These zones are located so that generally there is a capability to land at a secondary landing area once in every revolution.

Contingency Landing Area

The contingency landing area is that area outside the launch site, launch abort, primary, and secondary landing areas within which a landing could possibly occur, and requiring only the support of land-based contingency aircraft. For Apollo 9 this includes all the earth's surface between 34° North latitude and 34° South latitude (outside the areas mentioned above). Although there is a remote possibility that an immediate emergency or catastrophic failure could result in a landing anywhere within these latitudes, it is expected that most emergencies will permit sufficient time to delay the deorbit burn in order to land at or near a preselected target point.

The area is divided into four sectors for identification purposes:

1. Sector A (Atlantic Ocean)
2. Sector B (Indian Ocean)
3. Sector C (Western Pacific Ocean)
4. Sector D (Eastern Pacific Ocean)

No planned ship support of the contingency area is required; however, contingency aircraft deployed to various staging bases around the world are required. The aircraft will be located at the following staging bases to provide an 18-hour access time to any contingency landing; Hickam Air Force Base (AFB), Hawaii; Kindley AFB, Bermuda; Lajes Air Force Base (AFB) in the Azores; Tachikawa AB, Japan; Pago Pago, Samoa; Perth, Australia; Lima, Peru; the Ascension Islands; and the Mauritius Islands. Retrieval for a contingency area landing will be an after-the-fact operation.

Target Points

A preferred target point (PTP) will be selected each revolution to provide a landing opportunity at a desirable location. If possible, the PTP will be chosen in one of the four recovery zones supported by recovery ships and aircraft. If the ground track does not pass through a recovery zone, the PTP will be chosen near a recovery aircraft staging base.

Alternate Target Points (ATP's) will be selected approximately once every revolution in such a manner that they occur halfway between PTP's. As a result, there will be a PTP or an ATP approximately every 45 minutes during the flight. These preselected aiming points will be as close to search/rescue aircraft staging bases as practicable after considering such other things as weather, time of day of landing, and tracking capability.

MOBILE QUARANTINE FACILITY (MQF) SIMULATION OPERATIONS

During the Apollo 9 mission, NASA desires to exercise as realistically as practicable the Mobile Quarantine Facility (MQF) and its interfaces with the recovery forces. The MQF is a mobile living facility, 35 feet long, 9 feet wide, and 9 feet high (Figure 20). It is specially designed to biologically isolate a flight crew during the recovery phase of a lunar landing mission. For a lunar landing mission, the flight crew, one NASA flight surgeon, and one or two NASA support personnel will be biologically isolated in the MQF during its transportation from the recovery area to the Lunar Receiving Laboratory (LRL) at MSC. Although this exercise will be conducted concurrently with the Apollo 9 mission, it will in no way interfere with mission activities.

For the MQF exercise, a simulated CM (referred to as the "CM egress trainer") and associated equipment will be deployed aboard the USS Guadalcanal. The following procedures are planned to be accomplished while at sea. Two recovery simulations with the CM egress trainer will be required. The first one will take place approximately 24 to 48 hours after launch, and the second one between approximately 48 to 96 hours before the planned Apollo 9 recovery time. For these simulations, three NASA personnel simulating the flight crew will be inside the CM egress trainer and, before retrieval by helicopter, will don biological isolation garments. These personnel will then be retrieved and flown to the USS Guadalcanal where they will enter the MQF. The CM egress trainer will then be retrieved and mated to the MQF so that the removal of equipment can be simulated.

At this point in the second simulation, the normal routine planned for actual lunar landing missions (including obtaining blood samples and passing samples and equipment into and out of the MQF) will be followed. This routine will be continued through the Apollo 9 recovery and while the ship is enroute to Norfolk, Virginia.

At approximately 24 hours after recovery of Apollo 9, the USS Guadalcanal will arrive at Norfolk. The MQF and associated equipment will then be transferred to a C-141 aircraft and flown to MSC. Upon arrival at MSC, the MQF will be transported to the Lunar Receiving Laboratory (LRL) and a simulated docking and transfer into the laboratory performed. The operation will be terminated with an end-of-mission stowage exercise .

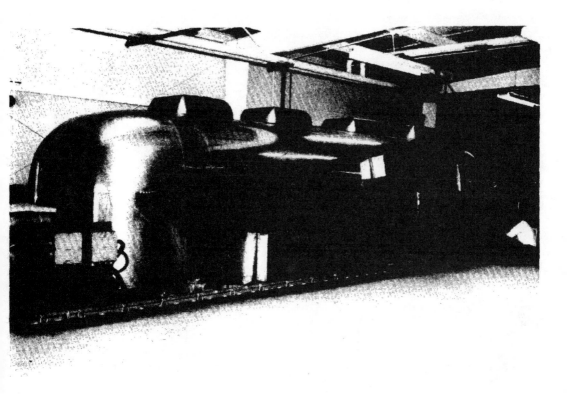

MOBILE QUARANTINE FACILITY   Fig. 20

FLIGHT CREW

FLIGHT CREW ASSIGNMENTS

Prime Crew (Figure 21)

Commander (CDR) - J. A. McDivitt (Colonel, USAF) Command Module Pilot (CMP) - D. R. Scott (Colonel, USAF) Lunar Module Pilot (LMP) - R. L. Schweickart (Civilian)

Backup Crew (Fig. 22)

Commander (CDR) - C. Conrad (Commander, USN) Command Module Pilot (CMP) - R. F. Gordon (Commander, USN) Lunar Module Pilot (LMP) - A. L. Bean (Lieutenant Commander, USN)

PRIME CREW BIOGRAPHICAL DATA

Fig. 21

Commander (CDR)

NAME: James A. McDivitt (Colonel, USAF)

DATE OF BIRTH: Born June 10, 1929, in Chicago, Illinois.

PHYSICAL DESCRIPTION-grown hair; blue eyes; height: 5 feet, 11 inches; weight, 155 pounds.

EDUCATION: Graduated from Kalamazoo Central High School, Kalamazoo, Michigan; received a Bachelor of Science degree in Aeronautical Engineering from the University of Michigan (graduated first in class) in 1959 and an Honorary Doctorate in Astronautical Science from the University of Michigan in 1965.

ORGANIZATIONS: Member of the Society of Experimental Test Pilots, the American Institute of Aeronautics and Astronautics, Tau Beta Pi, and Phi Kappa Phi.

SPECIAL HONORS: Awarded the NASA Exceptional Service Medal and the Air Force Astronaut Wings; four Distinguished Flying Crosses; Five Air Medals; the Chong Moo Medal from South Korea; the USAF Air Force Systems Command Aerospace Primus Award; the Arnold Air Society JFK Trophy; the Sword of Loyola; and the Michigan Wolverine Frontiersman Award.

EXPERIENCE: McDivitt joined the Air Force in 1951 and holds the rank of Colonel. He flew 145 combat missions during the Korean War in F-80's and F-86s. He is a graduate of the USAF Experimental Test Pilot School and the USAF Aerospace Research Pilot course and served as an experimental test pilot at Edwards Air Force Base, California.

CURRENT ASSIGNMENT: Colonel McDivitt was selected as an astronaut by NASA in September 1962.

He was command pilot for Gemini 4, a 66-orbit 4-day mission that began June 3 and ended on June 7, 1965. Highlights of the mission included a controlled extravehicular activity period performed by pilot Ed White, cabin depressurization and opening of spacecraft cabin doors, and the completion of 12 scientific and medical experiments.

<u>Command Module Pilot (CMP)</u>

NAME: David R. Scott (Colonel, USAF)

DATE OF BIRTH: Born June 6, 1932, in San Antonio, Texas.

PHYSICAL DESCRIPTION: Blond hair; blue eyes; height: 6 feet; weight: 175 pounds.

EDUCATION: Graduated from Western High School, Washington, D.C.; received a Bachelor of Science degree from the United States Military Academy and the degrees of Master of Science in Aeronautics and Astronautics and Engineer in Aeronautics and Astronautics from the Massachusetts Institute of Technology.

ORGANIZATIONS: Associate Fellow of the American Institute of Aeronautics and Astronautics; member of the Society of Experimental Test Pilots; Tau Beta Pi; Sigma Xi; and Sigma Gamma Tau.

SPECIAL HONORS: Awarded the NASA Exceptional Service Medal, the Air Force Astronaut Wings, and the Distinguished Flying Cross; and recipient of the AIAA Astronautics Award.

EXPERIENCE: Scott graduated fifth in a class of 633 at West Point and subsequently chose an Air Force career. He completed pilot training at Webb Air Force Base, Texas, in 1955.

He was assigned to the 32nd Tactical Fighter Squadron at Soesterberg Air Base (RNAF), Netherlands, from April 1956 to July 1960.

Upon completing this tour of duty, he returned to the United States for study at the Massachusetts Institute of Technology where he completed work on his Master's degree. His thesis at MIT concerned interplanetary navigation.

After completing his studies at MIT in June 1962, he attended the Air Force Experimental Test Pilot School and then the Aerospace Research Pilot School.

CURRENT ASSIGNMENT: Colonel Scott was one of the third group of astronauts named by NASA in October 1963. On March 16, 1966, he and Command Pilot Neil Armstrong were launched on the Gemini 8 mission flight originally scheduled to last three days but terminated early due to a malfunctioning OAMS thruster. The crew performed the first successful docking of two vehicles in space and demonstrated great piloting skill in overcoming the thruster problem and bringing the spacecraft to a safe landing.

Lunar Module Pilot (LMP)

NAME: Russell L. Schweickart (Civ.)

DATE OF BIRTH: Born October 25, 1935, in Neptune, New Jersey.

PHYSICAL DESCRIPTION: Red hair; blue eyes; height: 6 feet; weight: 161 pounds.

EDUCATION: Graduated from Manasquan High School, New Jersey; received a Bachelor of Science degree in Aeronautical Engineering and a Master of Science degree in Aeronautics and Astronautics from the Massachusetts Institute of Technology.

ORGANIZATIONS: Member of the Sigma Xi.

EXPERIENCE: Schweickart served as a pilot in the United States Air Force and Air National Guard from 1956 to 1963.
He was research scientist at the Experimental Astronomy Laboratory at MIT, and his work there included research in upper atmospheric physics, star tracking, and stabilization of stellar images. His thesis for a Master's degree at MIT concerned stratospheric radiance.

CURRENT ASSIGNMENT: Mr. Schweickart was one of the third group of astronauts named by NASA in October 1963.

BACKUP CREW BIOGRAPHICAL DATA

APOLLO 9 BACKUP CREW

CHARLES CONRAD, JR.   ALAN L. BEAN   RICHARD F. GORDON   Fig.

Commander (CDR)

NAME: Charles Conrad, Jr. (Commander, USN)

DATE OF BIRTH: Born on June 2, 1930, in Philadelphia, Pennsylvania.

PHYSICAL DESCRIPTION: Blond hair; blue eyes; height: 5 feet 6 1/2 inches; weight: 138 pounds.

EDUCATION: Attended primary and secondary schools in Haverford, Pennsylvania, and New Lebanon, New York; received a Bachelor of Science degree in Aeronautical Engineering from Princeton University in 1953 and an Honorary Master of Arts degree from Princeton in 1966.

ORGANIZATIONS: Member of the American Institute of Aeronautics and Astronautics and the Society of Experimental Test Pilots.

SPECIAL HONORS: Awarded two Distinguished Flying Crosses, two NASA Exceptional Service Medals, and the Navy Astronaut Wings; recipient of Princeton's Distinguished Alumnus Award for 1965, and the American Astronautical Society Flight Achievement Award for 1966.

EXPERIENCE: Conrad entered the Navy following his graduation from Princeton University and became a naval aviator. He attended the Navy Test Pilot School at Patuxent River, Maryland, and upon completing that course of instruction was assigned as a project test pilot in the armaments test division there. He also served at Patuxent as a flight instructor and performance engineer at the Test Pilot School.

CURRENT ASSIGNMENT: Commander Conrad was selected as an astronaut by NASA in September 1962. In August 1965, he served as Pilot on the 8-day Gemini 5 Flight. He and Command Pilot Gordon Cooper were launched on August 21 and proceeded to establish a new space endurance record of 190 hours and 56 minutes. The flight, which lasted 120 revolutions and covered a total distance of 3,312,993 statute miles, was terminated on August 29, 1965. It was also on this flight that the United States took over the lead in man hours in space.

On September 12, 1966, Conrad occupied the Command Pilot seat for the 3-day 44-revolution Gemini 11 mission. He executed orbital maneuvers to rendezvous and dock in less than one orbit with a previously launched Agena and controlled Gemini 11 through two periods of extravehicular activity performed by Pilot Richard Gordon.

Command Module Pilot (CMP)

NAME: Richard F. Gordon, Jr. (Commander, USN)

DATE OF BIRTH: Born October 5, 1929, in Seattle, Washington.

PHYSICAL DESCRIPTION: Brown hair; hazel eyes; height: 5 feet 7 inches; weight: 150 pounds.

EDUCATION: Graduated from North Kitsap High School, Poulsbo, Washington; received a Bachelor of Science degree in Chemistry from the University of Washington in 1951.

ORGANIZATIONS: Member of the Society of Experimental Test Pilots.

SPECIAL HONORS: Awarded two Distinguished Flying Crosses, the NASA Exceptional Service Medal, and the Navy Astronaut Wings.

EXPERIENCE: Gordon, a Navy Commander, received his wings as a naval aviator in 1953. He then attended All-Weather Flight School and jet transitional training and was subsequently assigned to an all-weather fighter squadron at the Naval Air Station at Jacksonville, Florida.

In 1957, he attended the Navy's Test Pilot School at Patuxent River, Maryland, and served as a flight test pilot until 1960.

He served with Fighter Squadron 121 at the Miramar, California, Naval Air Station as a flight instructor in the F4H and participated in the introduction of that aircraft to the Atlantic and Pacific Fleets. Winner of the

Bendix Trophy Race from Los Angeles to New York in May 1961, he established a new speed record of 869.
miles per hour and a transcontinental speed record of 2 hours and 47 minutes.

He was also a student at the U.S. Naval Postgraduate School at Monterey, California.

CURRENT ASSIGNMENT: Commander Gordon was one of the third group of astronauts named by NAS
in October 1963. He has since served as backup pilot for the Gemini 8 flight.

On September 12, 1966, he served as Pilot for the 3-day 44-revolution Gemini 11 mission on whic
rendezvous with an Agena was achieved in less than one orbit. He performed two periods of extravehicul
activity which included attaching a tether to the Agena and retrieving a nuclear emulsion experiment packag

Lunar Module Pilot (LMP)

NAME: Alan L. Bean (Lieutenant Commander, USN)

DATE OF BIRTH: Born in Wheeler, Texas, on March 15, 1932.

PHYSICAL DESCRIPTION: Brown hair; hazel eyes; height: 5 feet, 9 1/2 inches; weight, 155 pounds.

EDUCATION: Graduated from Paschal High School in Fort Worth, Texas; received a Bachelor of Scienc
degree in Aeronautical Engineering from the University of Texas in 1955.

ORGANIZATIONS: Member of the Society of Experimental Test Pilots and Delta Kappa Epsilon.

EXPERIENCE: Bean, a Navy ROTC student at Texas, was commissioned upon graduation in 1955. Upc
completing his flight training, he was assigned to Attack Squadron 44 at the Naval Air Station in Jacksonvill
Florida, for four years. He then attended the Navy Test Pilot School at Patuxent River, Maryland. Upc
graduation he was assigned as a test pilot at the Naval Air Test Center, Patuxent River. He attended the scho
of Aviation Safety at the University of Southern California and was next assigned to Attack Squadron 172
Cecil Field, Florida.

CURRENT ASSIGNMENT: Lt. Commander Bean was one of the third group of astronauts selected by NAS
in October 1963. He served as backup Command Pilot for the Gemini 10 mission.

MISSION MANAGEMENT RESPONSIBILITY

Title	Name	Organization
Director, Apollo Program	Lt. Gen. Sam C. Phillips	NASA/OMSF
Director, Mission Operations	Maj. Gen. John D. Stevenson (Ret)	NASA/OMSF
Saturn V Vehicle Prog. Mgr.	Mr. Lee B. James	NASA/MSFC
Apollo Spacecraft Prog. Mgr.	Mr. George M. Low	NASA/MSC
Apollo Prog. Manager KSC	R. Adm. Roderick O. Middleton	NASA/KSC
Mission Director	Mr. George H. Hage	NASA/OMSF
Assistant Mission Director	Capt. Chester M. Lee (Ret)	NASA/OMSF
Assistant Mission Director	Col. Thomas H. McMullen	NASA/OMSF
Director of Launch Operations	Mr. Rocco Petrone	NASA/KSC
Director of Flight Operations	Mr. Christopher C. Kraft	NASA/MSC
Launch Operations Manager	Mr. Paul C. Donnelly	NASA/KSC
Flight Directors	Mr. Eugene F. Kranz	
	Mr. Gerald D. Griffin	
	Mr. M. P. Frank	NASA/MSC
Spacecraft Commander (Prime)	Col. J. A. McDivitt	NASA/MSC
Spacecraft Commander (Backup)	Cdr. C. Conrad	NASA/MSC

PROGRAM MANAGEMENT

NASA HEADQUARTERS
Office of Manned Space Flight
Manned Spacecraft Center
Marshall Space Flight Center
Kennedy Space Center

LAUNCH VEHICLE
Marshall Space Flight Center
The Boeing Co. (S-IC)

North American Rockwell Corp. (S-II)
McDonnell Douglas Corp. (S-IVB)
IBM Corp. (IU)

SPACECRAFT
Manned Spacecraft Center
North American Rockwell
(LES, CSM, SLA)

Grumman Aircraft
Engineering Corp. (LM)

TRACKING AND DATA ACQUISITION
Kennedy Space Center
Goddard Space Flight Center

Department of Defense
MSFN

ABBREVIATIONS

AGS	Abort Guidance System
AK	Apogee Kick
APS	Ascent Propulsion System (LM)
APS	Auxiliary Propulsion System (S-IVB)
AS	Ascent Stage
ATP	Alternate Target Point
CCATS	Communications, Command, and Telemetry System
CDH	Constant Delta Height
CDR	Commander
CES	Central Electronics System
CFP	Concentric Flight Plan
CM	Command Module
CMP	Command Module Pilot
COI	Contingency Orbit Insertion
CSI	Concentric Sequence Initiation
CSM	Command Service Module
C/T	Crawler Transporter
DAP	Digital Auto Pilot
DOD	Department of Defense
DPS	Descent Propulsion System
DS	Descent Stage
DTO	Detailed Test Objectives
ECS	Environmental Control System
EDS	Emergency Detection System
EPS	Electrical Power System
EVA	Extravehicular Activity
GET	Ground Elapsed Time
G&N	Guidance and Navigation
GN&CS	Guidance, Navigation, and Control System
IMU	Inertial Measurement Unit
IU	Instrument Unit
IVT	Inter-vehicular Transfer
KSC	Kennedy Space Center
LC	Launch Complex
LCC	Launch Control Center

LES	Launch Escape System
LET	Launch Escape Tower
LH2	Liquid Hydrogen
LM	Lunar Module
LMP	Lunar Module Pilot
LOR	Lunar Orbit Rendezvous
LOX	Liquid Oxygen
LRL	Lunar Receiving Laboratory
LTA	Lunar Module Test Article
LV	Launch Vehicle
MCC	Mission Control Center
MOCR	Mission Operations Control Room
MQF	Mobile Quarantine Facility
MTVC	Manual Thrust Vector Control
OMSF	Office of Manned Space Flight
PGA	Pressure Garment Assembly
PGNCS	Primary Guidance, Navigation, and Control System
PTP	Preferred Target Point
RCS	Reaction Control System
RF	Radio Frequency
RSO	Range Safety Officer
RTCC	Real Time Computer Complex
S&A	Safe and Arm
SCS	Stabilization and Control System
SEQ	Sequential System
SLA	Spacecraft LM Adapter
SM	Service Module
SPS	Service Propulsion System
SV	Space Vehicle
SXT	Sextant
TB	Time Base
TD&E	Transposition, Docking, and Ejection
TLI	Trans Lunar Injection
TPI	Terminal Phase Initiation
VAB	Vehicle Assembly Building

MISSION OPERATION REPORT (APOLLO 9) SUPPLEMENT

25 FEBRUARY 1969

OFFICE OF MANNED SPACEFLIGHT
Prepared by: Apollo Program Office - MAO

SPACE VEHICLE

The primary flight hardware of the Apollo program consists of a Saturn V Launch Vehicle and an Apollo Spacecraft. Collectively, they are designated the Apollo-Saturn V Space Vehicle (SV) (Figure 1).

SATURN V LAUNCH VEHICLE

The Saturn V Launch Vehicle (LV) is designed to boost up to 285,000 pounds into a 105 nautical mile earth orbit and to provide for lunar payloads of 100,000 pounds. The Saturn V LV consists of three propulsive stages (S-IC, S-II, S-IVB), two interstages, and an Instrument Unit (IU).

S-IC Stage

General

The S-IC stage (Figure 2) is a large cylindrical booster, 138 feet long and 33 feet in diameter, powered by five liquid propellant F-1 rocket engines. These engines develop a nominal sea level thrust total of approximately 7,650,000 pounds and have an operational burn time of 159 seconds. The stage dry weight is approximately 295,300 pounds and the total loaded stage weight is approximately 5,031,500 pounds. The S-IC stage interfaces structurally and electrically with the S-II stage. It also interfaces structurally, electrically, and pneumatically with Ground Support Equipment (GSE) through two umbilical service arms, three tail service masts, and certain electronic systems by antennas. The S-IC stage is instrumented for operational measurements or signals which are transmitted by its independent telemetry system.

Structure

The S-IC structural design reflects the requirements of F-1 engines, propellants, control, instrumentation, and interfacing systems. Aluminum alloy is the primary structural material .The major structural components are the forward skirt, oxidizer tank, intertank section, fuel tank, and thrust structure. The forward skirt interfaces structurally with the S-IC/S-II interstage. The skirt also mounts vents, antennas, electrical and electronic equipment.

The 47,298-cubic foot oxidizer tank is the structural link between the forward skirt and the intertank structure which provides structural continuity between the oxidizer and fuel tanks. The 29,215-cubic foot fuel tank provides the load carrying structural link between the thrust and intertank structures. Five oxidizer ducts run from the oxidizer tank, through the fuel tank, to the F-1 engines.

The thrust structure assembly redistributes the applied loads of the five F-1 engines into nearly uniform loading about the periphery of the fuel tank. Also, it provides support for the five F-1 engines, engine accessories, base heat shield, engine fairings and fins, propellant lines, retrorockets, and environmental control ducts. The lower thrust ring has four holddown points which support the fully loaded Saturn V Space Vehicle (approximately 6,483,000 pounds) and also, as necessary, restrain the vehicle during controlled release.

Propulsion

The F-1 engine is a single-start, 1,530,000-pound fixed-thrust, calibrated, bi-propellant engine which uses Liquid Oxygen (LOX) as the oxidizer and Rocket Propellant-1 (RP-1) as the fuel. The thrust chamber is cooled regeneratively by fuel, and the nozzle extension is cooled by gas generator exhaust gases. Oxidizer and fuel are supplied to the thrust chamber by a single turbopump powered by a gas generator which uses the same propellant combination. RP-1 is also used as the turbopump lubricant and as the working fluid for the engine hydraulic control system. The four outboard engines are capable of gimbaling and have provisions for supply and return of RP-1 as the working fluid for a thrust vector control system. The engine contains a heat exchanger system to condition engine-supplied LOX and externally supplied helium for stage propellant tank pressurization. An instrumentation system monitors engine performance and operation. External thermal insulation provides an allowable engine environment during flight operation.

APOLLO-SATURN V SPACE VEHICLE

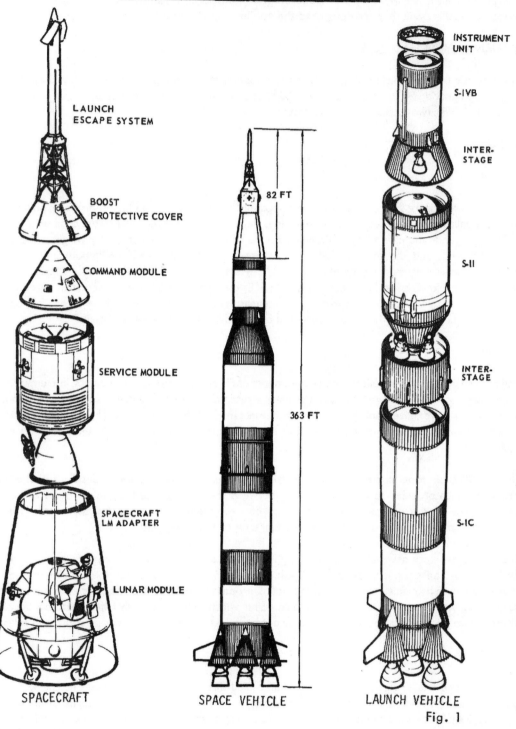

SPACECRAFT

LAUNCH ESCAPE SYSTEM

BOOST PROTECTIVE COVER

COMMAND MODULE

SERVICE MODULE

SPACECRAFT LM ADAPTER

LUNAR MODULE

SPACE VEHICLE

82 FT

363 FT

LAUNCH VEHICLE

INSTRUMENT UNIT

S-IVB

INTER-STAGE

S-II

INTER-STAGE

S-IC

Fig. 1

The normal in-flight engine cutoff sequence is center engine first, followed by the four outboard engines. Engine optical-type depletion sensors in either the oxidizer or fuel tank initiate the engine cutoff sequence. In an emergency, the engine can be cut off by any of the following methods: Ground Support Equipment (GSE) Command Cutoff, Emergency Detection System, or Outboard Cutoff System.

S-IC STAGE

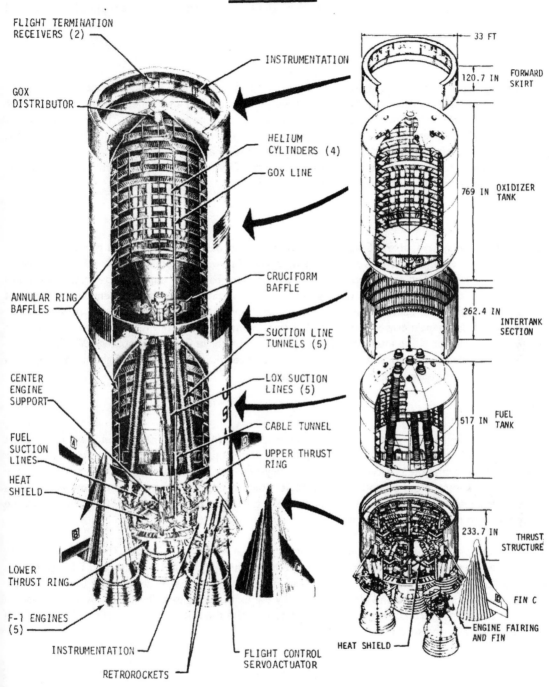

FLIGHT TERMINATION
RECEIVERS (2)

INSTRUMENTATION

GOX
DISTRIBUTOR

HELIUM
CYLINDERS (4)

GOX LINE

ANNULAR RING
BAFFLES

CRUCIFORM
BAFFLE

SUCTION LINE
TUNNELS (5)

CENTER
ENGINE
SUPPORT

LOX SUCTION
LINES (5)

CABLE TUNNEL

FUEL
SUCTION
LINES

UPPER THRUST
RING

HEAT
SHIELD

LOWER
THRUST RING

F-1 ENGINES
(5)

INSTRUMENTATION

RETROROCKETS

FLIGHT CONTROL
SERVOACTUATOR

33 FT

120.7 IN   FORWARD
            SKIRT

769 IN   OXIDIZER
          TANK

262.4 IN   INTERTANK
            SECTION

517 IN   FUEL
          TANK

233.7 IN   THRUST
            STRUCTURE

FIN C

ENGINE FAIRING
AND FIN

HEAT SHIELD

Fig. 2

Propellant Systems

The propellant systems include hardware for fill and drain, propellant conditioning, and tank pressurization prior to and during flight, and for delivery to the engines. Fuel tank pressurization is required during engine starting and flight to establish and maintain a Net Positive Suction Head (NPSH) at the fuel inlet to the engine turbopumps. During flight, the source of fuel tank pressurization is helium from storage bottles mounted inside the oxidizer tank. Fuel feed is accomplished through two 12-inch ducts which connect the fuel tank to each F-1 engine. The ducts are equipped with flex and sliding joints to compensate for motions from engine gimbaling and stage stresses.

Gaseous Oxygen (GOX) is used for oxidizer tank pressurization during flight. A portion of the LOX supplied to each engine is diverted into the engine heat exchangers where it is transformed into GOX and routed back to the tanks. LOX is delivered to the engines through five suction lines which are supplied with flex and sliding joints.

Flight Control System

The S-IC thrust vector control consists of four outboard F-1 engines, gimbal blocks to attach these engines to the thrust ring, engine hydraulic servoactuators (two per engine), and an engine hydraulic power supply. Engine thrust is transmitted to the thrust structure through the engine gimbal block. There are two servo-actuator attach points per engine, located 90 degrees from each other, through which the gimbaling force is applied. The gimbaling of the four outboard engines changes the direction of thrust and as a result corrects the attitude of the vehicle to achieve the desired trajectory. Each outboard engine may be gimbaled ±5° within a square pattern at a rate of 5° per second.

Electrical

The electrical power system of the S-IC stage consists of two basic subsystems: the operational power subsystem and the measurements power subsystem. Onboard power is supplied by two 28-volt batteries. Battery number 1 is identified as the operational power system battery. It supplies power to operational loads such as valve controls, purge and venting systems, pressurization systems, and sequencing and flight control. Battery number 2 is identified as the measurement power system. Batteries supply power to their loads through a common main power distributor, but each system is completely isolated from the other. The S-IC stage switch selector is the interface between the Launch Vehicle Digital Computer (LVDC) in the IU and the S-IC stage electrical circuits. Its function is to sequence and control various flight activities such as telemetry calibration, retrofire initiation, and pressurization.

Ordnance

The S-IC ordnance systems include the propellant dispersion (flight termination) and the retrorocket systems. The S-IC Propellant Dispersion System (PDS) provides the means of terminating the flight of the Saturn V if it varies beyond the prescribed limits of its flight path or if it becomes a safety hazard during the S-IC boost phase. A transmitted ground command shuts down all engines and a second command detonates explosives which longitudinally open the fuel and oxidizer tanks. The fuel opening is 180° (opposite) to the oxidizer opening to minimize propellant mixing.

Eight retrorockets provide thrust after S-IC burnout to separate it from the S-II stage. The S-IC retrorockets are mounted in pairs external to the thrust structure in the fairings of the four outboard F-1 engines. The firing command originates in the IU and activates redundant firing systems. At retrorocket ignition the forward end of the fairing is burned and blown through by the exhausting gases. The thrust level developed by seven retrorockets (one retrorocket out) is adequate to separate the S-IC stage a minimum of six feet from the vehicle in less than one second.

S-II Stage

General

The S-II stage (Figure 3) is a large cylindrical booster, 81.5 feet long and 33 feet in diameter, powered by five liquid propellant J-2 rocket engines which develop a nominal vacuum thrust of 230,000 pounds each for a total of 1,150,000 pounds. Dry weight of the S-II stage is approximately 84,600 pounds. The stage approximate loaded gross weight is 1,064,600 pounds. The S-IC/S-II interstage weighs 11,665 pounds. The S-II stage is instrumented for operational and R&D measurements which are transmitted by its independent telemetry system. The S-II stage has structural and electrical interfaces with the S-IC and S-IV stages, and electric, pneumatic, and fluid interfaces with GSE through its umbilicals and antennas.

Structure

Major S-II structural components are the forward skirt, the 37,737-cubic foot fuel tank, the 12,745 -cubic foot oxidizer tank (with the common bulkhead), the aft skirt/thrust structure, and the S-IC/S-II interstage. Aluminum alloy is the major structural material. The forward and aft skirts distribute and transmit structural loads and interface structurally with the interstages. The aft skirt also distributes the loads imposed on the thrust structure by the J-2 engines. The S-IC/S-II interstage is comparable to the aft skirt in capability and construction. The propellant tank walls constitute the cylindrical structure between the skirts. The aft bulkhead of the fuel tank is also the forward bulkhead of the oxidizer tank. This common bulkhead is fabricated of aluminum with a fiberglass/ phenolic honeycomb core. The insulating characteristics of the common bulkhead minimize the heating effect of the relatively hot LOX (-297°F) on the LH_2 (-423°F).

Propulsion

The S-II stage engine system consists of five single-start, high-performance, high altitude J-2 rocket engines of 230,000 pounds of nominal vacuum thrust each. Fuel is liquid hydrogen (LH_2) and the oxidizer is liquid oxygen (LOX). The four outer J-2 engines are equally spaced on a 17.5-foot diameter circle and are capable of being gimbaled through a ±7 degree square pattern to allow thrust vector control. The fifth engine is fixed and is mounted on the centerline of the stage. The engine valves are controlled by a pneumatic system powered by gaseous helium which is stored in a sphere inside the start tank. An electrical control system that uses solid state logic elements is used to sequence the start and shutdown operations of the engine. Electrical power is stage-supplied.

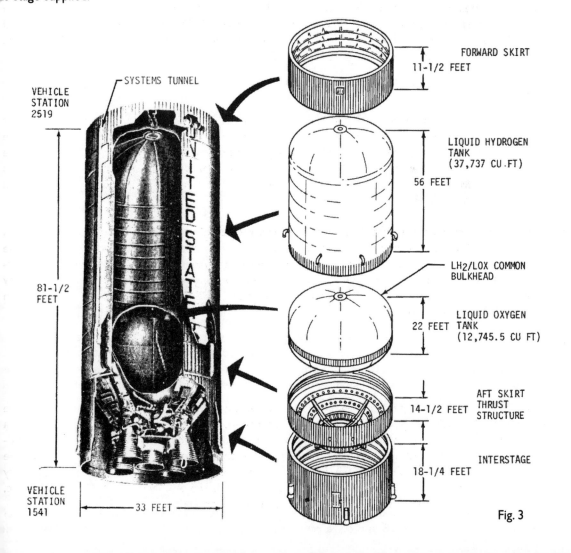

Fig. 3

The J-2 engine may receive cutoff signals from several different sources. These sources include engine interlock deviations, EDS automatic or manual abort cutoffs, and propellant depletion cutoff. Each of these sources signals the LVDC in the IU. The LVDC sends the engine cutoff signal to the S-II switch selector, which in turn signals the electrical control package, which controls all local signals necessary for the cutoff sequence. Five discrete liquid level sensors per propellant tank provide initiation of engine cutoff upon detection of propellant depletion. The cutoff sensors will initiate a signal to shut down the engines when two out of five engine cutoff signals from the same tank are received.

Propellant Systems

The propellant systems supply fuel and oxidizer to the five engines. This is accomplished by the propellant management components and the servicing, conditioning, and engine delivery subsystems. The propellant tanks are insulated with foam-filled honeycomb which contains passages through which helium is forced for purging and leak detection. The LH_2 feed system includes five 8-inch vacuum jacketed feed ducts and five prevalves.

During powered flight, prior to S-II ignition, Gaseous Hydrogen (GH_2) for LH_2 tank pressurization is bled from the thrust chamber hydrogen injector manifold of each of the four outboard engines. After S-II engine ignition, LH_2 is preheated in the regenerative cooling tubes of the engine and topped off from the thrust chamber injector manifold in the form of GH_2 to serve as a pressurizing medium. The LOX feed system includes four 8-inch, vacuum-jacketed feed ducts, one uninsulated feed duct, and five prevalves. LOX tank pressurization is accomplished with GOX obtained by heating LOX bled from the LOX turbopump outlet.

The propellant management system monitors propellant mass for control of propellant loading, utilization and depletion. Components of the system include continuous capacitance probes, propellant utilization valves, liquid level sensors, and electronic equipment. During flight, the signals from the tank continuous capacitance probes are monitored and compared to provide an error signal to the propellant utilization valve on each LOX pump. Based on this error signal, the propellant utilization valves are positioned to minimize residual propellants and assure a fuel-rich cutoff by varying the amount of LOX delivered to the engines.

Flight Control System

Each outboard engine is equipped with a separate, independent, closed-loop, hydraulic control system that includes two servoactuators mounted in perpendicular planes to provide vehicle control in pitch, roll, and yaw. The servoactuators are capable of deflecting the engine ±7 degrees in the pitch and yaw planes (±10 degrees diagonally) at the rate of 8 degrees per second.

Electrical

The electrical system is comprised of the electrical power and electrical control subsystems. The electrical power system provides the S-II stage with the electrical power source and distribution. The electrical control system interfaces with the IU to accomplish the mission requirements of the stage. The LVDC in the IU controls in-flight sequencing of stage functions through the stage switch selector. The stage switch selector outputs are routed through the stage electrical sequence controller or the separation controller to accomplish the directed operation. These units are basically a network of low-power transistorized switches that can be controlled individually and, upon command from the switch selector, provide properly sequenced electrical signals to control the stage functions.

Ordnance

The S-II ordnance systems include the separation, ullage rocket, retrorocket, and propellant dispersion (flight termination) systems. For S-IC/S-II separation, a dual-plane separation technique is used wherein the structure between the two stages is severed at two different planes. The second-plane separation jettisons the interstage after S-II engine ignition. The S-II/S-IVB separation occurs at a single plane located near the aft skirt of the S-IVB stage. The S-IVB interstage remains as an integral part of the S-II stage. To separate and retard the S-II stage, a deceleration is provided by the four retrorockets located in the S-II/S-IVB interstage. Each rocket develops a nominal thrust of 34,810 pounds and fires for 1.52 seconds. All separations are initiated by the LVDC located in the IU.

To ensure stable flow of propellants into the J-2 engines, a small forward acceleration is required to settle the propellants in their tanks. This acceleration is provided by four ullage rockets mounted on the S-IC interstage. Each rocket develops a nominal thrust of 23,000 pounds and fires for 3.75 seconds. The ullage function occurs prior to second-plane separation.

The S-II Propellant Dispersion System (PDS) provides for termination of vehicle flight during the S-II boost phase if the vehicle flight path varies beyond its prescribed limits or if continuation of vehicle flight creates a safety hazard. The S-II PDS may be safed after the launch escape tower is jettisoned. The fuel tank linear shaped charge, when detonated, cuts a 30-foot vertical opening in the tank. The oxidizer tank destruct charges simultaneously cut 13-foot lateral openings in the oxidizer tank and the S-II aft skirt.

S-IVB Stage

General

The S-IVB stage (Figure 4) is a large cylindrical booster 59 feet long and 21.6 feet in diameter, powered by one J-2 engine. The S-IVB stage is capable of multiple engine starts. Engine thrust is 232,000 pounds for the first burn and 206,000 pounds for subsequent burns. This stage is also unique in that it has an attitude control capability independent of its main engine. Dry weight of the stage is 25,300 pounds. The launch weight of the stage is 259,160 pounds. The interstage weight of 8,080 pounds is not included in the stated weights. The stage is instrumented for functional measurements or signals which are transmitted by its independent telemetry system.

Structure

The major structural components of the S-IVB stage are the forward skirt, propellant tanks, aft skirt, thrust structure and aft interstage. The forward skirt provides structural continuity between the fuel tank walls and the IU. The propellant tank walls transmit and distribute structural loads from the aft skirt and the thrust structure. The aft skirt is subjected to imposed loads from the S-IVB aft interstage. The thrust structure mounts the J-2 engine and distributes its structural loads to the circumference of the oxidizer tank. A common, insulated bulkhead separates the 2830-cubic foot oxidizer tank and the 10,418-cubic foot fuel tank and is similar to the common bulkhead discussed in the S-II description. The predominant structural material of the stage is aluminum alloy. The stage interfaces structurally with the S-II stage and the IU.

Main Propulsion

The high-performance 232,000-pound thrust J-2 engine as installed in the S-IVB stage has a multiple restart capability. The engine valves are controlled by a pneumatic system powered by gaseous helium which is stored in a sphere inside a start bottle. An electrical control system that uses solid state logic elements is used to sequence the start and shutdown operations of the engine. Electrical power is supplied from aft battery No. 1.

During engine operation, the oxidizer tank is pressurized by flowing cold helium (from helium spheres mounted inside the fuel tank) through the heat exchanger in the oxidizer turbine exhaust duct. The heat exchanger heats the cold helium, causing it to expand. The fuel tank is pressurized during engine operation by GH_2 from the thrust chamber fuel manifold. Thrust vector control in the pitch and yaw planes during burn periods is achieved by gimbaling the entire engine.

The J-2 engine may receive cutoff signals from the following sources; EDS, range safety systems, "Thrust OK" pressure switches, propellant depletion sensors, and an IU programmed command (velocity or timed) via the switch selector.

S-IVB STAGE

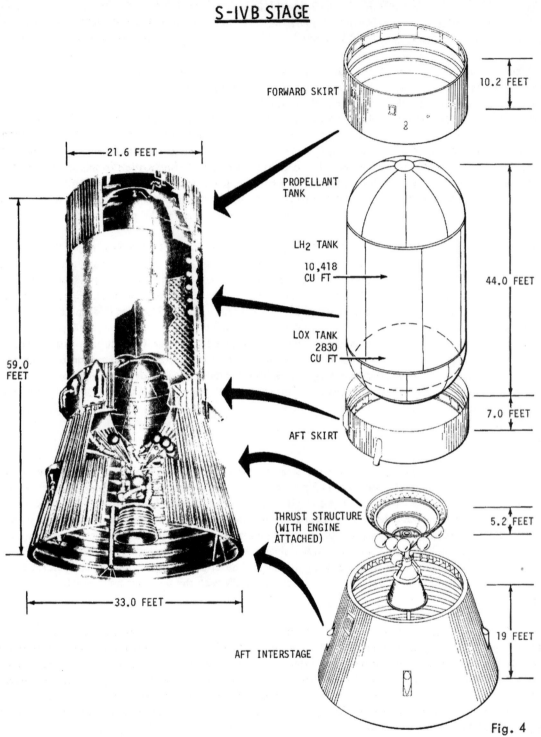

FORWARD SKIRT

10.2 FEET

21.6 FEET

PROPELLANT
TANK

LH₂ TANK

10,418
CU FT

44.0 FEET

LOX TANK
2830
CU FT

7.0 FEET

AFT SKIRT

59.0
FEET

THRUST STRUCTURE
(WITH ENGINE
ATTACHED)

5.2 FEET

33.0 FEET

19 FEET

AFT INTERSTAGE

Fig. 4

The restart of the J-2 engine is identical to the initial start except for the fill procedure of the start tank. The start tank is filled with LH_2 and GH_2 during the first burn period by bleeding GH_2 from the thrust chamber fuel injection manifold and LH_2 from the Augmented Spark Igniter (ASI) fuel line to refill the start tank for engine restart. (Approximately 50 seconds of mainstage engine operation is required to recharge the start tank.)

To insure that sufficient energy will be available for spinning the fuel and oxidizer pump turbines, a waiting period of between approximately 80 minutes to 6 hours is required. The minimum time is required to build sufficient pressure by warming the start tank through natural means and to allow the hot gas turbine exhaust ystem to cool. Prolonged heating will cause a loss of energy in the start tank. This loss occurs when the LH_2 nd GH_2 warm and raise the gas pressure to the relief valve setting. If this venting continues over a prolonged period the total stored energy will be depleted. This limits the waiting period prior to a restart attempt to ix hours.

Propellant Systems

LOX is stored in the aft tank of the propellant tank structure at a temperature of -297°F. A six-inch, ow-pressure supply duct supplies LOX from the tank to the engine. During engine burn, LOX is supplied at nominal flow rate of 392 pounds per second, and at a transfer pressure above 25 psia. The supply duct is equipped with bellows to provide compensating flexibility for engine gimbaling, manufacturing tolerances, and hermal movement of structural connections. The tank is prepressurized to between 38 and 41 psia and is maintained at that pressure during boost and engine operation. Gaseous helium is used as the pressurizing gent.

The LH_2 is stored in an insulated tank at less than -423°F. LH_2 from the tank is supplied to the J-2 engine urbopump by a vacuum-jacketed, low-pressure, 110-inch duct. This duct is capable of flowing 80 pounds per econd at -423°F and at a transfer pressure of 28 psia. The duct is located in the aft tank side wall above the ommon bulkhead joint. Bellows in this duct compensate for engine gimbaling, manufacturing tolerances, and hermal motion. The fuel tank is prepressurized to 28 psia minimum and 31 psia maximum.

The PU subsystem provides a means of controlling the propellant mass ratio. It consists of oxidizer and fuel ank mass probes, a PU valve, and an electronic assembly. These components monitor the propellant and maintain command control. Propellant utilization is provided by bypassing oxidizer from the oxidizer turbo-pump outlet back to the inlet. The PU valve is controlled by signals from the PU system. The engine oxidizer/fuel mixture mass ratio varies from 4.5:1 to 5.5:1.

Flight Control System

The flight control system incorporates two systems for flight and attitude control. During powered flight, hrust vector steering is accomplished by gimbaling the J-2 engine for pitch and yaw control and by operating he Auxiliary Propulsion System (APS) engines for roll control. The engine is gimbaled in a ±7.5 degree square pattern by a closed-loop hydraulic system. Mechanical feedback from the actuator to the servovalve provides he closed engine position loop. Two actuators are used to translate the steering signals into vector forces o position the engine. The deflection rates are proportional to the pitch and yaw steering signals from the ight control computer. Steering during coast flight is by use of the APS engine alone.

Auxiliary Propulsion System

The S-IVB APS provides three-axis stage attitude control (Figure 5) and main stage propellant control during oast flight. The APS engines are located in two modules 180° apart on the aft skirt of the S-IVB stage (Figure). Each module contains four engines; three 150-pound thrust control engines, and one 70-pound thrust llage engine. Each module contains its own oxidizer, fuel, and pressurization system. A positive expulsion ropellant feed subsystem is used to assure that hypergolic propellants are supplied to the engines under zero g" or random gravity conditions. Nitrogen tetroxide (N_2O_4), is the oxidizer and monomethyl hydrazine MMH), is the fuel for these engines.

Electrical

The electrical system of the S-IVB stage is comprised of two major subsystems: the electrical power ubsystem which consists of all the power sources on the stage; and the electrical control subsystem which istributes power and control signals to various loads throughout the stage. Onboard electrical power is upplied by four silver-zinc batteries. Two are located in the forward equipment area and two in the aft quipment area. These batteries are activated and installed in the stage during the final pre-launch reparations. Heaters and instrumentation probes are an integral part of each battery.

APS FUNCTIONS

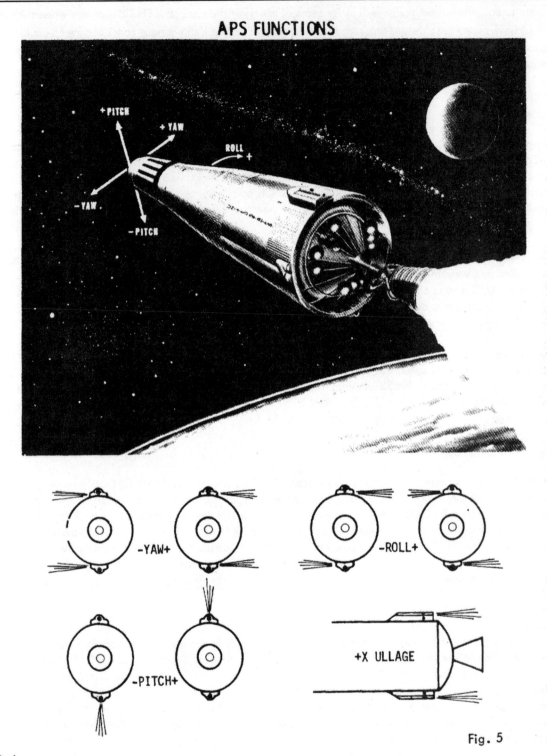

Fig. 5

Ordnance

The S-IVB ordnance systems include the separation, ullage rocket, and PDS systems. The separation plane for S-II/S-IVB staging is located at the top of the S-II/S-IVB interstage. At separation four retrorocket motors mounted on the interstage structure below the separation plane fire to decelerate the S-II stage with the interstage attached.

To provide propellant settling and thus ensure stable flow of fuel and oxidizer during J-2 engine start, the S-IVB stage requires a small acceleration. This acceleration is provided by two jettisonable ullage rockets for the first burn. The APS provides ullage for subsequent burns.

The S-IVB PDS provides for termination of vehicle flight by cutting two parallel 20-foot openings in the fuel tank and a 47-inch diameter hole in the LOX tank. The S-IVB PDS may be safed after the launch escape tower is jettisoned. Following S-IVB engine cutoff at orbit insertion, the PDS is electrically safed by ground command.

Instrument Unit

General

The Instrument Unit (IU) (Figure 7), is a cylindrical structure 21.6 feet in diameter and 3 feet high installed on top of the S-IVB stage. The IU contains the guidance, navigation, and control equipment for the Launch Vehicle. In addition, it contains measurements and telemetry, command communications, tracking, and emergency detection system components along with supporting electrical power and environmental control systems.

Structure

The basic IU structure is a short cylinder fabricated of an aluminum alloy honeycomb sandwich material. Attached to the inner surface of the cylinder are "cold plates" which serve both as mounting structure and thermal conditioning units for the electrical/electronic equipment.

Navigation, Guidance, and Control

The Saturn V Launch Vehicle is guided from its launch pad into earth orbit by navigation, guidance, and control equipment located in the IU. An all-inertial system utilizes a space-stabilized platform for acceleration and attitude measurements. A Launch Vehicle Digital Computer (LVDC) is used to solve guidance equations and a Flight Control Computer (FCC) (analog) is used for the flight control functions.

The three-gimbal stabilized platform (ST-124-M3) provides a space-fixed coordinate reference frame for attitude control and for navigation (acceleration) measurements. Three integrating accelerometers, mounted on the gyro-stabilized inner gimbal of the platform, measure the three components of velocity resulting from vehicle propulsion. The accelerometer measurements are sent through the Launch Vehicle Data Adapter (LVDA) to the LVDC. In the LVDC, the accelerometer measurements are combined with the computed gravitational acceleration to obtain velocity and position of the vehicle. During orbital flight, the navigational program continually computes the vehicle position, velocity, and acceleration. Guidance information stored in the LVDC (e.g., position, velocity) can be updated through the IU command system by data transmission

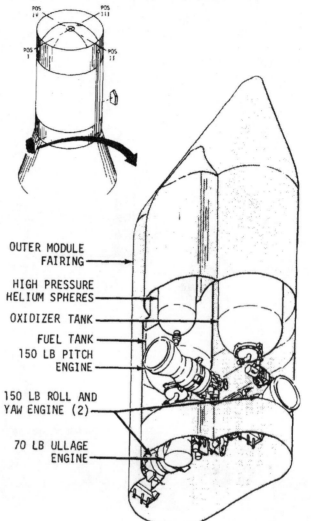

APS CONTROL MODULE

OUTER MODULE FAIRING

HIGH PRESSURE HELIUM SPHERES

OXIDIZER TANK

FUEL TANK

150 LB PITCH ENGINE

150 LB ROLL AND YAW ENGINE (2)

70 LB ULLAGE ENGINE

Fig. 6

SATURN INSTRUMENT UNIT

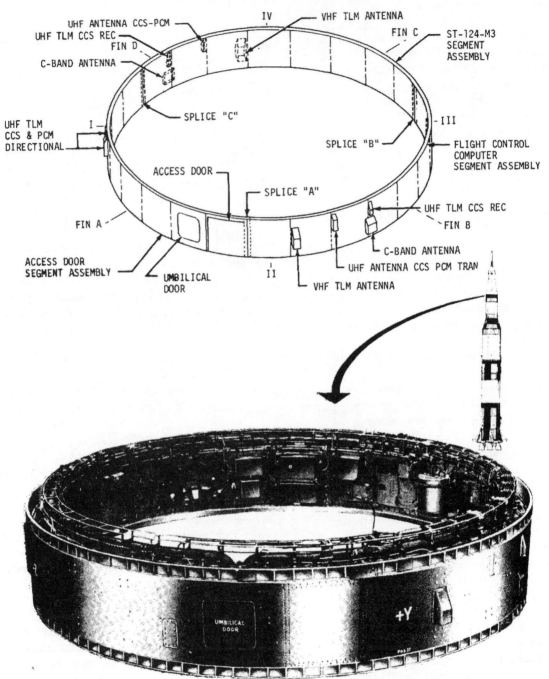

Fig. 7

from ground stations. The IU command system provides the general capability of changing or inserting information into the LVDC.

The control subsystem is designed to maintain and control vehicle attitude by forming the steering commands to be used by the controlling engines of the active stage. The control system accepts guidance computations from the LVDC/LVDA Guidance System. These guidance commands, which are actually attitude error signals, are then combined with measured data from the various control sensors. The resultant output is the command signal to the various engine actuators and APS nozzles. The final computations (analog) are performed within the FCC. The FCC is also the central switching point for command signals. From this point, the signals are routed to their associated active stages and to the appropriate attitude control devices.

Measurements and Telemetry

The instrumentation within the IU consists of a measuring subsystem, a telemetry subsystem and an antenna subsystem. This instrumentation is for the purpose of monitoring certain conditions and events which take place within the IU and for transmitting monitored signals to ground receiving stations.

Command Communications System

The Command Communications System (CCS) provides for digital data transmission from ground stations to the LVDC. This communications link is used to update guidance information or command certain other functions through the LVDC. Command data originates in the Mission Control Center (MCC) and is sent to remote stations of the Manned Space Flight Network (MSFN) for transmission to the Launch Vehicle.

Saturn Tracking Instrumentation

The Saturn V IU carries two C-band radar transponders and an Azusa/GLOTRAC tracking transponder. A combination of tracking data from different tracking systems provides the best possible trajectory information and increased reliability through redundant data. The tracking of the Saturn Launch Vehicle may be divided into four phases: powered flight into earth orbit; orbital flight; injection into mission trajectory; and coast flight after injection. Continuous tracking is required during powered flight into earth orbit. During orbital flight, tracking is accomplished by S-band stations of the MSFN and by C-band radar stations.

IU Emergency Detection System Components

The Emergency Detection System (EDS) is one element of several crew safety systems. There are nine EDS rate gyros installed in the IU. Three gyros monitor each of the three axes (pitch, roll, and yaw) thus providing triple redundancy. The control signal processor provides power to and receives inputs from the nine EDS rate gyros. These inputs are processed and sent on to the EDS distributor and to the flight control computer. The EDS distributor serves as a junction box and switching device to furnish the spacecraft display panels with emergency signals if emergency conditions exist. It also contains relay and diode logic for the automatic abort sequence.

An electronic timer in the IU allows multiple engine shutdowns without automatic abort after 30 seconds of flight. Inhibiting of automatic abort circuitry is also provided by the vehicle flight sequencing circuits through the IU switch selector. This inhibiting is required prior to normal S-IC engine cutoff and other normal vehicle sequencing. While the automatic abort is inhibited, the flight crew must initiate a manual abort if an angular overrate or two engine-out condition arises.

Electrical Power Systems

Primary flight power for the IU equipment is supplied by silver-zinc oxide batteries at a nominal voltage level of 28 ±2 vdc. Where ac power is required within the IU it is developed by solid state dc to ac inverters. Power distribution within the IU is accomplished through power distributors which are essentially junction boxes and switching circuits.

Environmental Control System

The Environmental Control System (ECS) maintains an acceptable operating environment for the IU equipment during pre-flight and flight operations. The ECS is composed of the following:

I. The Thermal Conditioning System (TCS) which maintains a circulating coolant temperature to the electronic equipment of 59° ±1°F.

2. Pre-flight purging system which maintains a supply of temperature and pressure regulated air/gaseous nitrogen in the IU/S-IVB equipment area.

3. Gas bearing supply system which furnishes gaseous nitrogen to the ST-124-M3 inertial platform gas bearings.

4. Hazardous gas detection sampling equipment which monitors the IU/S-IVB forward interstage area for the presence of hazardous vapors.

APOLLO SPACECRAFT

The Apollo Spacecraft (S/C) is designed to support three men in space for periods up to two weeks, docking in space, landing on and returning from the lunar surface, and safely reentering the earth's atmosphere. The Apollo S/C consists of the Spacecraft LM Adapter (SLA), the Service Module (SM), the Command Module (CM), Launch Escape System (LES), and the Lunar Module (LM).

Spacecraft LM Adapter

General

The SLA (Figure 8) is a conical structure which provides a structural load path between the LV and SM and also supports the LM. Aerodynamically, the SLA smoothly encloses the irregularly shaped LM and transitions the space vehicle diameter from that of the upper stage of the LV to that of the SM. The SLA also encloses the nozzle of the SM engine and the High Gain Antenna.

Structure

The SLA is constructed of 1.7 inch thick aluminum honeycomb panels. The four upper jettisonable, or forward, panels are about 21 feet long, and the fixed lower, or aft, panels about 7 feet long. The exterior surface of the SLA is covered completely by a layer of cork. The cork helps insulate the LM from aerodynamic heating during boost. The LM is attached to the SLA at four locations around the lower panels.

SLA-SM Separation

The SLA and SM are bolted together through flanges on each of the two structures. Explosive trains are used to separate the SLA and SM as well as for separating the four upper jettisonable SLA panels. Redundancy is provided in three areas to assure separation; redundant initiating signals, redundant detonators and cord trains, and "sympathetic" detonation of nearby charges.

Pyrotechnic type and spring type thrusters (Figure 9) are used in deploying and jettisoning the SLA upper panels. The four double-piston pyrotechnic thrusters are located inside the SLA and start the panels swinging outward on their hinges. The two pistons of the thruster push on the ends of adjacent panels thus providing two separate thrusters operating each panel. The explosive train which separates the panels is routed through two pressure cartridges in each thruster assembly. The pyrotechnic thrusters rotate the panels 2 degrees establishing a constant angular velocity of 33 to 60 degrees per second. When the panels have rotated about 45 degrees, the partial hinges disengage and free the panels from the aft section of the SLA, subjecting them to the force of the spring thrusters.

The spring thrusters are mounted on the outside of the upper panels. When the panel hinges disengage, the springs in the thruster push against the fixed lower panels to propel the panels away from the vehicle at an angle of 110 degrees to the centerline at a speed of about 5½ miles per hour. The panels will then depart the area of the spacecraft.

SPACECRAFT LM ADAPTER

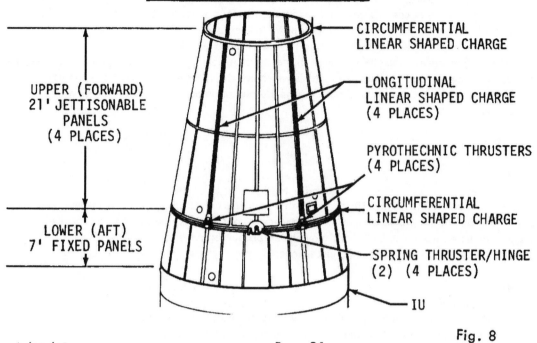

UPPER (FORWARD)
21' JETTISONABLE
PANELS
(4 PLACES)

CIRCUMFERENTIAL
LINEAR SHAPED CHARGE

LONGITUDINAL
LINEAR SHAPED CHARGE
(4 PLACES)

PYROTHECHNIC THRUSTERS
(4 PLACES)

CIRCUMFERENTIAL
LINEAR SHAPED CHARGE

LOWER (AFT)
7' FIXED PANELS

SPRING THRUSTER/HINGE
(2) (4 PLACES)

IU

Fig. 8

SLA PANEL JETTISONING

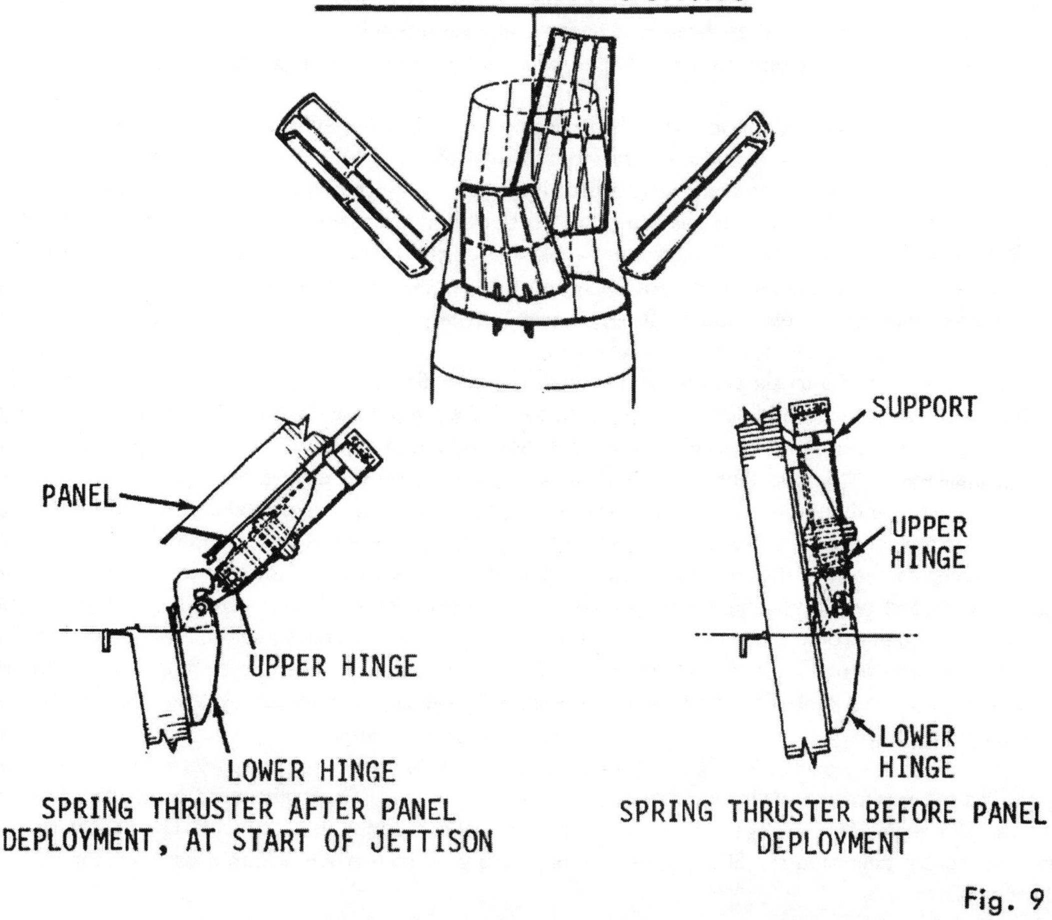

PANEL

SUPPORT

UPPER HINGE

UPPER HINGE

LOWER HINGE

LOWER
HINGE

SPRING THRUSTER AFTER PANEL
DEPLOYMENT, AT START OF JETTISON

SPRING THRUSTER BEFORE PANEL
DEPLOYMENT

Fig. 9

SLA-LM Separation

Spring thrusters also are used to separate the LM from the SLA. After the command/ service module has docked with the LM, mild charges are fired to release the four adapters which secure the LM in the SLA. Simultaneously, four spring thrusters mounted on the lower (fixed) SLA panels push against the LM Landing Gear Truss Assembly to separate the spacecraft from the launch vehicle. The separation is controlled by two LM Separation Sequence Controllers located inside the SLA near the attachment point to the Instrument Unit (IU). The redundant controllers send signals which fire the charges that sever the connections and also fire a detonator to cut the LM-IU Umbilical. The detonator impels a guillotine blade which severs the umbilical wires.

Service Module

General

The Service Module (SM) (Figure 10) provides the main spacecraft propulsion and maneuvering capability during a mission. The SM provides most of the spacecraft consumables (oxygen, water, propellant, hydrogen) and supplements environmental, electrical power, and propulsion requirements of the CM. The SM remains attached to the CM until it is jettisoned just before CM reentry.

Structure

The basic structural components are forward and aft (upper and lower) bulkheads, six radial beams, four sector honeycomb panels, four reaction control system honeycomb panels, aft heat shield, and a fairing. The forward and aft bulkheads cover the top and bottom of the SM. Radial beam trusses extending above the forward bulkhead support and secure the CM. The radial beams are made of solid aluminum alloy which has been machined and chem-milled to thickness' varying between 2 inches and 0.018 inch. Three of these beams have compression pads and the other three have shear-compression pads and tension ties. Explosive charges in the center sections of these tension ties are used to separate the CM from the SM.

An aft heat shield surrounds the service propulsion engine to protect the SM from the engine's heat during thrusting. The gap between the CM and the forward bulkhead of the SM is closed off with a fairing which is composed of eight electrical power system radiators alternated with eight aluminum honeycomb panels. The sector and reaction control system panels are one inch thick and are made of aluminum honeycomb core between two aluminum face sheets. The sector panels are bolted to the radial beams. Radiators used to dissipate heat from the environmental control subsystem are bonded to the sector panels on opposite sides of the SM. These radiators are each about 30 square feet in area.

The SM interior is divided into six sectors and a center section. Sector one is currently void. It is available for installation of scientific or additional equipment should the need arise. Sector two has part of a space radiator and an RCS engine quad (module) on its exterior panel and contains the SPS oxidizer sump tank. This tank is the larger of the two tanks that hold the oxidizer for the SPS engine. Sector three has the rest of the space radiator and another RCS engine quad on its exterior panel and contains the oxidizer storage tank. This tank is the second of two SPS oxidizer tanks and is fed from the oxidizer sump tank in sector two. Sector four contains most of the electrical power generating equipment. It contains three fuel cells, two cryogenic oxygen and two cryogenic hydrogen tanks and a power control relay box. The cryogenic tanks supply oxygen to the environmental control subsystem and oxygen and hydrogen to the fuel cells. Sector five has part of an environmental control radiator and an RCS engine quad on the exterior panel and contains the SPS engine fuel sump tank. This tank feeds the engine and is also connected by feed lines to the fuel storage tank in sector six.

Sector six has the rest of the environmental control radiator and an RCS engine quad on its exterior and contains the SPS engine fuel storage tank which feeds the fuel sump tank in sector five. The center section contains two helium tanks and the SPS engine. The tanks are used to provide helium pressurant for the SPS propellant tanks.

Propulsion

Main spacecraft propulsion is provided by the 20,500-pound thrust Service Propulsion System (SPS). The SPS engine is a restartable, non-throttleable engine which uses nitrogen tetroxide as an oxidizer and a 50-50 mixture of hydrazine and unsymmetrical dimethylhydrazine as fuel. This engine is used for major velocity changes during the mission such as midcourse corrections, lunar orbit insertion, trans-earth injection, and CSM aborts. The SPS engine responds to automatic firing commands from the guidance and navigation system or to commands from manual controls. The engine assembly is gimbal-mounted to allow engine thrust-vector alignment with the spacecraft center of mass to preclude tumbling. Thrust vector alignment control is maintained automatically by the stabilization and control system or manually by the crew. The Service Module Reaction Control System (SM RCS) provides for maneuvering about and along three axes. (See page 179 for more comprehensive description.)

Additional SM Systems

In addition to the systems already described the SM has communication antennas, umbilical connections, and several exterior mounted lights. The four antennas on the outside of the SM are the steerable S-band high-gain antenna, mounted on the aft bulkhead; two VHF omni directional antennas, mounted on opposite sides of the module near the top; and the rendezvous radar transponder antenna, mounted in the SM fairing.

The umbilicals consist of the main plumbing and wiring connections between the CM and SM enclosed in a fairing (aluminum covering), and a "flyaway" umbilical which is connected to the launch tower. The latter supplies oxygen and nitrogen for cabin pressurization, water-glycol, electrical power from ground equipment, and purge gas.

Seven lights are mounted in the aluminum panels of the fairing. Four (one red, one green, and two amber) are used to aid the astronauts in docking, one is a floodlight which can be turned on to give astronauts visibility during extravehicular activities, one is a flashing beacon used to aid in rendezvous, and one is a spotlight used in rendezvous from 500 feet to docking with the LM.

SM/CM Separation

Separation of the SM from the CM occurs shortly before reentry. The sequence of events during separation is controlled automatically by two redundant Service Module Jettison Controllers (SMJC) located on the forward bulkhead of the SM. Physical separation requires severing of all the connections between the modules, transfer of electrical control, and firing of the SM RCS to increase the distance between the CM and SM. A tenth of a second after electrical connections are deadfaced, the SMJC's send signals which fire ordnance devices to sever the three tension ties and the umbilical. The tension ties are straps which hold the CM on three of the compression pods on the SM. Linear-shaped charges in each tension tie assembly sever the tension ties to separate the CM from the SM. At the same time, explosive charges drive guillotines through the wiring and tubing in the umbilical. Simultaneously with the firing of the ordnance devices, the SMJC's send signals which fire the SM RCS. Roll engines are fired for five seconds to alter the SM's course from that of the CM, and the translation (thrust) engines are fired continuously until the propellant is depleted or fuel cell power is expended. These maneuvers carry the SM well away from the entry path of the CM.

Command Module

General

The Command Module (CM) (Figure 11) serves as the command, control, and communications center for most of the mission. Supplemented by the SM, it provides all life support elements for three crewmen in the mission environments and for their safe return to earth's surface. It is capable of attitude control about three axes and some lateral lift translation at high velocities in earth atmosphere. It also permits LM attachment, CM/LM ingress and egress, and serves as a buoyant vessel in open ocean.

Structure

The CM consists of two basic structures joined together: the inner structure (pressure shell) and the outer structure (heat shield). The inner structure, the pressurized crew compartment, is made of aluminum sandwich construction consisting of a welded aluminum inner skin, bonded aluminum honeycomb core and outer face sheet. The outer structure is basically a heat shield and is made of stainless steel brazed honeycomb brazed between steel alloy face sheets. Parts of the area between the inner and outer sheets is

COMMAND MODULE

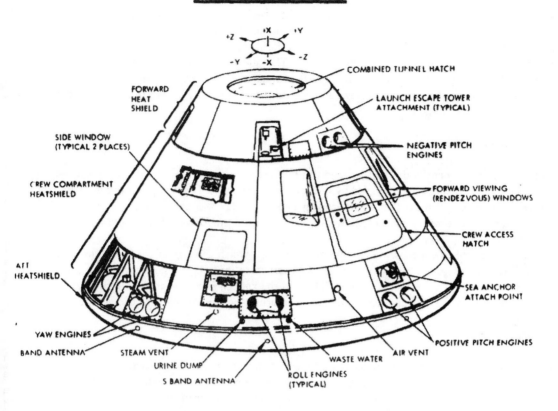

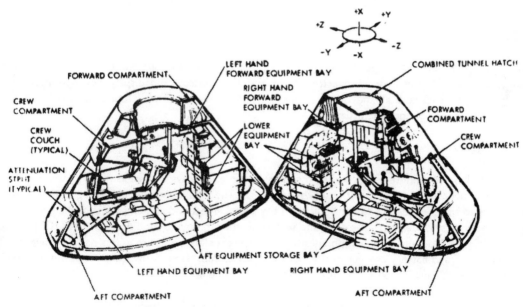

Fig. 11

filled with a layer of fibrous insulation as additional heat protection.

Thermal Protection (Heat Shields)

The interior of the CM must be protected from the extremes of environment that will be encountered during a mission. The heat of launch is absorbed principally through the Boost Protective Cover (BPC), a fiberglass structure covered with cork which encloses the CM. The cork is covered with a white reflective coating. The BPC is permanently attached to the launch escape tower and is jettisoned with it.

The insulation between the inner and outer shells, plus temperature control provided by the environmental control subsystem, protects the crew and sensitive equipment in space. The principal task of the heat shield that forms the outer structure is to protect the crew during reentry. This protection is provided by ablative heat shields of varying thickness' covering the CM. The ablative material is a phenolic epoxy resin. This material turns white hot, chars, and then melts away, conducting relatively little heat to the inner structure. The heat shield has several outer coverings: a pore seal, a moisture barrier (white reflective coating), and a silver Mylar thermal coating.

Forward Compartment

The forward compartment is the area around the forward (docking) tunnel. It is separated from the crew compartment by a bulkhead and covered by the forward heat shield. The compartment is divided into four 90-degree segments which contain earth landing equipment (all the parachutes, recovery antennas and beacon light, and sea recovery sling, etc.), two RCS engines, and the forward heat shield release mechanism.

The forward heat shield contains four recessed fittings into which the legs of the launch escape tower are attached. The tower legs are connected to the CM structure by frangible nuts containing small explosive charges, which separate the tower from the CM when the Launch Escape System is jettisoned. The forward heat shield is jettisoned at about 25,000 feet during return to permit deployment of the parachutes.

Aft Compartment

The aft compartment is located around the periphery of the CM at its widest part, near the aft heat shield. The aft compartment bays contain 10 RCS engines; the fuel, oxidizer, and helium tanks for the CM RCS; water tanks; the crushable ribs of the impact attenuation system; and a number of instruments. The CM-SM umbilical is also located in the aft compartment.

Crew Compartment

The crew compartment has a habitable volume of 210 cubic feet. Pressurization and temperature are maintained by the ECS. The crew compartment contains the controls and displays for operation of the spacecraft, crew couches, and all the other equipment needed by the crew. It contains two hatches, five windows, and a number of equipment bays.

Equipment Bays

The equipment bays contain items needed by the crew for up to 14 days, as well as much of the electronics and other equipment needed for operation of the spacecraft. The bays are named according to their position with reference to the couches. The lower equipment bay is the largest and contains most of the guidance and navigation electronics, as well as the sextant and telescope, the Command Module Computer (CMC), and a computer keyboard. Most of the telecommunications subsystem electronics are in this bay, including the five batteries, inverters, and battery charger of the electrical power subsystem. Stowage areas in the bay contain food supplies, scientific instruments, and other astronaut equipment.

The left-hand equipment bay contains key elements of the ECS. Space is provided in this bay for stowing the forward hatch when the CM and LM are docked and the tunnel between the modules is open. The left-hand forward equipment bay also contains ECS equipment, as well as the water delivery unit and clothing storage.

The right-hand equipment bay contains waste management system controls and equipment, electrical power equipment, and a variety of electronics, including sequence controllers and signal conditioners. Food also is stored in a compartment in this bay. The right-hand forward equipment bay is used principally for stowage and contains such items as survival kits, medical supplies, optical equipment, the LM docking target, and bioinstrumentation harness equipment.

The aft equipment bay is used for storing space suits and helmets, life vests, the fecal canister, portable life support systems (backpacks), and other equipment, and includes space for stowing the probe and drogue assembly.

Hatches

The two CM hatches are the side hatch, used for getting in and out of the CM, and the forward hatch, used to transfer to and from the LM when the CM and LM are docked. The side hatch is a single integrated assembly which opens outward and has primary and secondary thermal seals. The hatch normally contains a small window, but has provisions for installation of an airlock. The latches for the side hatch are so designed that pressure exerted against the hatch serves only to increase the locking pressure of the latches. The hatch handle mechanism also operates a mechanism which opens the access hatch in the BPC. A counterbalance assembly which consists of two nitrogen bottles and a piston assembly enables the hatch and BPC hatch to be opened easily. In space, the crew can operate the hatch easily without the counter balance, and the piston cylinder and nitrogen bottle can be vented after launch. A second nitrogen bottle can be used to open the hatch after landing. The side hatch can readily be opened from the outside. In case some deformation or other malfunction prevented the latches from engaging, three jackscrews are provided in the crew's tool set to hold the door closed.

The forward (docking) hatch is a combined pressure and ablative hatch mounted at the top of the docking tunnel. The exterior or upper side of the hatch is covered with a half-inch of insulation and a layer of aluminum foil. This hatch has a six point latching arrangement operated by a pump handle similar to that on the side hatch and can also be opened from the outside. It has a pressure equalization valve so that the pressure in the tunnel and that in the LM can be equalized before the hatch is removed. There are also provisions for opening the latches manually if the handle gear mechanism should fail.

Windows

The CM has five windows: two side (numbers 1 and 5), two rendezvous (numbers 2 and 4), and a hatch window (number 3 or center). The hatch window is over the center couch. The windows each consist of inner and outer panes. The inner windows are made of tempered silica glass with quarter-inch thick double panes, separated by a tenth of an inch. The outer windows are made of amorphous-fused silicon with a single pane seven tenths of an inch thick. Each pane has an anti-reflecting coating on the external surface and a blue-red reflective coating on the inner surface to filter out most infrared and all ultraviolet rays. The outer window glass has a softening temperature of 2800°F and a melting point of 3110°F. The inner window glass has a softening temperature of 2000°F. Aluminum shades are provided for all windows.

Impact Attenuation

During a water impact the CM deceleration force will vary considerably depending on the shape of the waves and the dynamics of the CM's descent. A major portion of the energy (75 to 90 percent) is absorbed by the water and by deformation of the CM structure. The impact attenuation system reduces the forces acting on the crew to a tolerable level. The impact attenuation system is part internal and part external. The external part consists of four crushable ribs (each about four inches thick and a foot in length) installed in the aft compartment. The ribs are made of bonded laminations of corrugated aluminum which absorb energy by collapsing upon impact. The main parachutes suspend the CM at such an angle that the ribs are the first point of the module that hits the water. The internal portion of the system consists of eight struts which connect the crew couches to the CM structure. These struts absorb energy by deforming steel wire rings between an inner and an outer piston.

Displays and Controls

The Main Display Console (Figure 12) has been arranged to provide for the expected duties of crew

MAIN DISPLAY CONSOLE

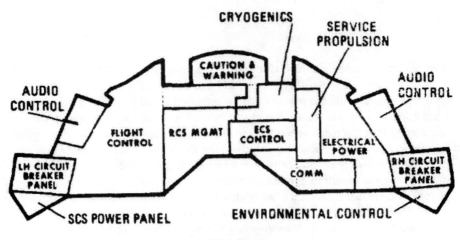

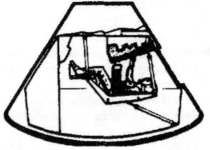

- LAUNCH VEHICLE EMERGENCY DETECTION
- FLIGHT ATTITUDE
- MISSION SEQUENCE
- VELOCITY CHANGE MONITOR
- ENTRY MONITOR

- PROPELLANT GAUGING
- ENVIRONMENT CONTROL
- COMMUNICATIONS CONTROL
- POWER DISTRIBUTION
- CAUTION & WARNING

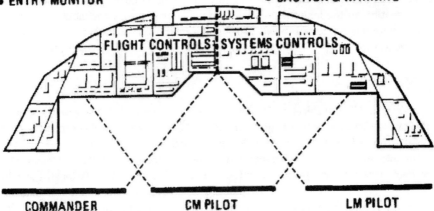

Fig. 12

members. These duties fall into the categories of Commander, CM Pilot, and LM Pilot, occupying the left, center, and right couches respectively. The CM Pilot also acts as the principal navigator. All controls have been designed so they can be operated by astronauts wearing gloves. The controls are predominantly of four basic types: toggle switches, rotary switches with click-stops, thumb wheels, and push buttons. Critical switches are guarded so that they cannot be thrown inadvertently. In addition, some critical controls have locks that must be released before they can be operated.

Flight controls are located on the left-center and left side of the Main Display Console, opposite the Commander. These include controls for such subsystem, as stabilization and control, propulsion, crew safety, earth landing, and emergency detection. One of two guidance and navigation computer panels also is located here, as are velocity, attitude, and altitude indicators.

The CM Pilot faces the center of the console and thus can reach many of the flight controls, as well as the system controls on the right side of the console. Displays and controls directly opposite him include reaction control propellant management, caution and warning, environmental control, and cryogenic storage systems.

The LM Pilot couch faces the right-center and right side of the console. Communications, electrical control, data storage, and fuel cell system components are located here, as well as service propulsion subsystem propellant management.

Other displays and controls are placed throughout the cabin in the various equipment bays and on the crew couches. Most of the guidance and navigation equipment is in the lower equipment bay, at the foot of the center couch. This equipment, including the sextant and telescope, is operated by an astronaut standing and using a simple restraint system. The non-time-critical controls of the environmental control system are located in the left-hand equipment bay, while all the controls of the waste management system are on a panel in the right-hand equipment bay. The rotation and translation controllers used for attitude, thrust vector, and translation maneuvers are located on the arms of two crew couches. In addition, a rotation controller can be mounted at the navigation position in the lower equipment bay.

Critical conditions of most spacecraft systems are monitored by a Caution And Warning System. A malfunction or out-of-tolerance condition results in illumination of a status light that identifies the abnormality. It also activates the master alarm circuit, which illuminates two master alarm lights on the Main Display Console and one in the lower equipment bay and sends an alarm tone to the astronauts' headsets. The master alarm lights and tone continue until a crewman resets the master alarm circuit. This can be done before the crewmen deal with the problem indicated. The Caution And Warning System also contains equipment to sense its own malfunctions.

Telecommunications

The telecommunications system (Figure 13) provides voice, television, telemetry, tracking, and ranging communications between the spacecraft and earth, between the CM and LM, and between the spacecraft and astronauts wearing the Portable Life Support System (PLSS). It also provides communications among the astronauts in the spacecraft and includes the central timing equipment for synchronization of other equipment and correlation of telemetry equipment. For convenience, the telecommunications subsystem can be divided into four areas: intercommunications (voice), data, radio frequency equipment, and antennas.

Intercommunications

The astronauts' headsets are used for all voice communications. Each astronaut has an audio control panel on the Main Display Console which enables him to control what comes into his headset and where he will send his voice. The three headsets and audio control panels are connected to three identical audio center modules. The audio center is the assimilation and distribution point for all spacecraft voice signals. The audio signals can be routed from the center to the appropriate transmitter or receiver, the Launch Control Center (for pre-launch checkout), the recovery forces intercom, or voice tape recorders.

Two methods of voice transmission and reception are possible: The VHF/AM transmitter-receiver and the S-band transmitter and receiver. The VHF/AM equipment is used for voice communications with the Manned Space Flight Network during launch, ascent, and near-earth phases of a mission. The S-band equipment is used during both near-earth and deep-space phases of a mission. When communications with earth are not possible, a limited number of audio signals can be stored on tape for later transmission. The CSM communication range capability is depicted in Figure 14.

TELECOMMUNICATIONS SYSTEM

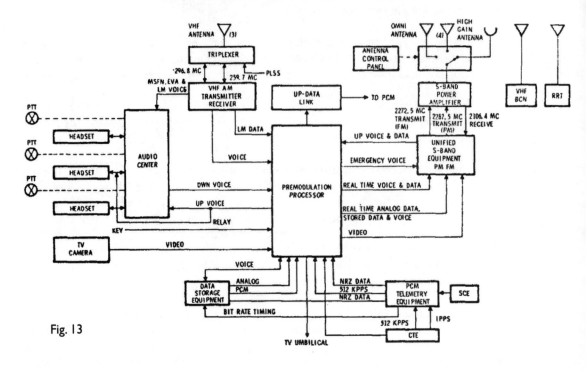

Fig. 13

Data

The spacecraft structure and subsystems contain sensors which gather data on their status and performance. Biomedical, TV, and timing data also are gathered. These various forms of data are assimilated into the data system, processed, and then transmitted to the ground. Some data from the operational systems, and some voice communications, may be stored for later transmission or for recovery after landing. Stored data can be transmitted to the ground simultaneously with voice or realtime data.

Radio Frequency Equipment

The radio frequency equipment is the means by which voice information, telemetry data, and ranging and tracking information are transmitted and received. The equipment consists of two VHF/AM transceivers in one unit, the unified S-band equipment (primary and secondary transponders and an FM transmitter), primary and secondary S-band power amplifiers (in one unit), a VHF beacon, an X-band transponder (for rendezvous radar), and the premodulation processor.

The equipment provides for voice transfer between the CM and the ground, between the CM and LM, between the CM and extravehicular astronauts, and between the CM and recovery forces. Telemetry can be transferred between the CM and the ground, from the LM to the CM and then to the ground, and from extravehicular astronauts to the CM and then to the ground. Ranging information consists of pseudo-random noise and double-Doppler ranging signals from the ground to the CM and back to the ground, and of X-band radar signals from the LM to the CM and back to the LM. The VHF beacon equipment emits a 2-second signal every five seconds for line-of-sight direction finding to aid recovery forces in locating the CM after landing.

Antennas

There are nine antennas (Figure 15) on the CSM, not counting the rendezvous radar antenna which is an integral part of the rendezvous radar transponder. These antennas can be divided into four groups: VHF, S-band, recovery, and beacon. The two VHF antennas (called scimitars because of their shape) are omni

CSM COMMUNICATION RANGES

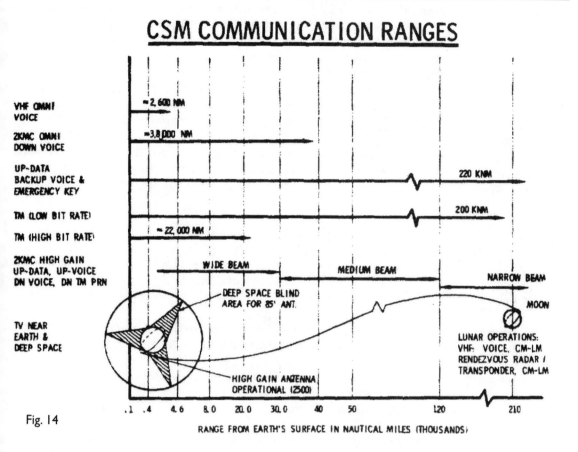

Fig. 14

RANGE FROM EARTH'S SURFACE IN NAUTICAL MILES (THOUSANDS)

LOCATION OF ANTENNAS

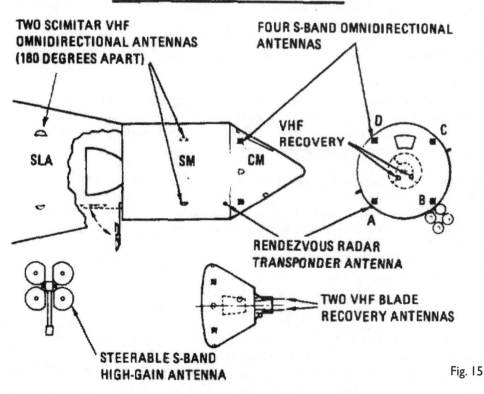

Fig. 15

directional and are mounted 180 degrees apart on the SM. There are five S-band antennas, one mounted at the bottom of the SM and four located 90 degrees apart around the CM. The S-band high-gain antenna, used for deep space communications, is composed of four 31-inch diameter reflectors surrounding an 11-inch square reflector. At launch it is folded down parallel to the SPS engine nozzle so that it fits within the spacecraft LM adapter. After the CSM separates from the SLA the antenna is deployed at a right angle to the SM center line. The four smaller surface-mounted S-band antennas are used at near-earth ranges and deep-space backup. The high-gain antenna is deployable after CSM/SLA separation. It can be steered through a gimbal system and is the principal antenna for deep-space communications. The four S-band antennas on the CM are mounted flush with the surface of the CM and are used for S-band communications during near-earth phases of the mission, as well as for a backup in deep space. The two VHF recovery antennas are located in the forward compartment of the CM, and are deployed automatically shortly after the main parachutes deploy. One of these antennas also is connected to the VHF recovery beacon.

Environmental Control System

The Environmental Control System (ECS) provides a controlled environment for three astronauts for up to 14 days. For normal conditions, this environment includes a pressurized cabin (five pounds per square inch), a 100-percent oxygen atmosphere, and a cabin temperature of 70 to 75 degrees Fahrenheit. The system provides a pressurized suit circuit for use during critical mission phases and for emergencies.

The ECS provides oxygen and hot and cold water, removes carbon dioxide and odors from the CM cabin, provides for venting of waste, and dissipates excessive heat from the cabin and from operating electronic equipment. It is designed so that a minimum amount of crew time is needed for its normal operation. The main unit contains the coolant control panel, water chiller, two water-glycol evaporators, carbon dioxide odor-absorber canisters, suit heat exchanger, water separator, and compressors. The oxygen surge tank, water glycol pump package and reservoir, and control panels for oxygen and water are adjacent to the unit.

The system is concerned with three major elements: oxygen, water, and coolant (water-glycol). All three are interrelated and intermingled with other systems. These three elements provide the major functions of spacecraft atmosphere, thermal control, and water management through four major subsystems: oxygen, pressure suit circuit, water, and water-glycol. A fifth subsystem, post-landing ventilation, also is part of the environmental control system, providing outside air for breathing and cooling prior to hatch opening.

The CM cabin atmosphere is 60 percent oxygen and 40 percent nitrogen on the launch pad to reduce fire hazard. The mixed atmosphere supplied by ground equipment will gradually be changed to pure oxygen after launch as the environmental control system maintains pressure and replenishes the cabin atmosphere.

During pre-launch and initial orbital operation, the suit circuit supplies pure oxygen at a flow rate slightly more than is needed for breathing and suit leakage. This results in the suit being pressurized slightly above cabin pressure, which prevents cabin gases from entering and contaminating the suit circuit. The excess oxygen in the suit circuit is vented into the cabin.

Spacecraft heating and cooling is performed through two water-glycol coolant loops. The water-glycol, initially cooled through ground equipment, is pumped through the primary loop to cool operating electric and electronic equipment, the space suits, and the cabin heat exchangers. The water-glycol also is circulated through a reservoir in the CM to provide a heat sink during ascent.

Earth Landing System

The Earth Landing System (ELS) (Figure 16) provides a safe landing for the astronauts and the CM. Several recovery aids which are activated after splashdown are part of the system. Operation normally is automatic, timed, and activated by the sequential control system. All automatic functions can be backed up manually.

For normal entry, about 1.5 seconds after forward heat shield jettison, the two drogue parachutes are deployed to orient the CM properly and to provide initial deceleration. At about 10,000 feet, the drogue

ELS MAJOR COMPONENT STOWAGE

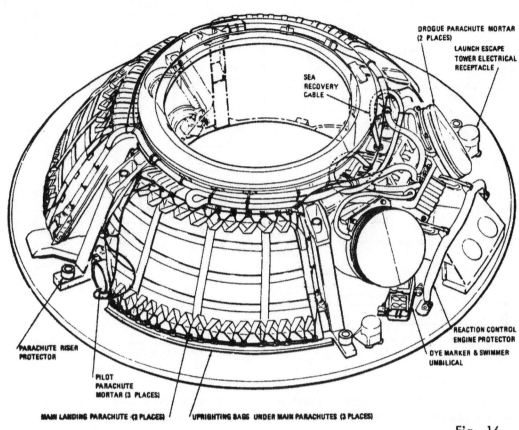

Fig. 16

parachutes are released and the three pilot parachutes are deployed; these pull the main parachutes from the forward section of the CM. The main parachutes initially open partially (reefed) for ten seconds to limit deceleration prior to full-diameter deployment. The main parachutes hang the CM at an angle of 27.5 degrees to decrease impact loads at touchdown.

After splashdown the crew releases the main parachutes and sets the recovery aid subsystem in operation. The subsystem consists of an uprighting system, swimmer's umbilical cable, a sea dye marker, a flashing beacon, and a VHF beacon transmitter. A sea recovery sling of steel cable is provided to lift the CM aboard a recovery ship. Three inflatable uprighting bags, stowed under the main parachutes, are available for uprighting the CM should it stabilize in an inverted floating position after splashdown.

The two VHF recovery antennas are located in the forward compartment with the parachutes. They are deployed automatically eight seconds after the main parachutes. One of them is connected to the beacon transmitter which emits a two second signal every five seconds to aid recovery forces in locating the CM. The other is connected to the VHF/AM transmitter and receiver to provide voice communications between the crew and recovery forces.

Common Spacecraft Systems

Guidance and Control

The Apollo spacecraft is guided and controlled by two interrelated systems (Figure 17). One is the Guidance, Navigation, and Control System (GNCS). The other is the Stabilization and Control System (SCS). The two

GUIDANCE AND CONTROL FUNCTIONAL FLOW

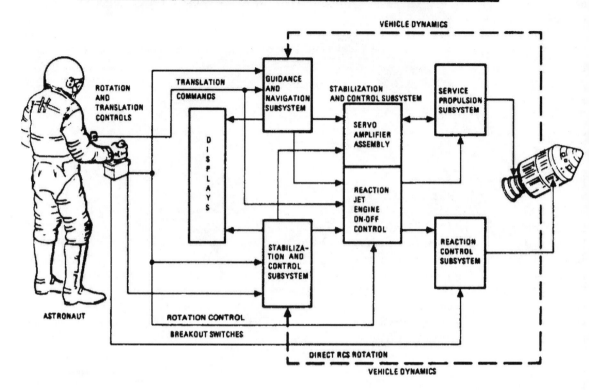

Fig. 17

systems provide rotational, line-of-flight, and rate-of-speed information. They integrate and interpret this information and convert it into commands for the spacecraft propulsion systems.

Guidance, Navigation, and Control System

Guidance and navigation is accomplished through three major elements; the inertial, optical, and computer systems. The inertial subsystem senses any changes in the velocity and angle of the spacecraft and relays this information to the computer which transmits any necessary signals to the spacecraft engines. The optical subsystem is used to obtain navigation sightings of celestial bodies and landmarks on the earth and moon. It passes this information along to the computer for guidance and control purposes. The computer subsystem uses information from a number of sources to determine the spacecraft position and speed and, in automatic operation, to give commands for guidance and control.

Stabilization and Control System

The Stabilization and Control System (SCS) operates in three ways; it determines the spacecraft's attitude (angular position); maintains the spacecraft's attitude; and controls the direction of thrust of the service propulsion engine. Both the GNCS and SCS are used by the computer in the CM to provide automatic control of the Spacecraft. Manual control of the spacecraft attitude and thrust is provided mainly through the SCS equipment.

The Flight Director Attitude Indicators (FDAI) on the main console show the total angular position, attitude errors, and their rates of change. One of the sources of total attitude information is the stable platform of the Inertial Measurement Unit (IMU). The second source is a Gyro Display Coupler (GDC) which gives a reading of the spacecraft's actual attitudes as compared with an attitude selected by the crew. Information about attitude error also is obtained by comparison of the IMU gimbal angles with computer reference angles. Another source of this information is gyro assembly No. I, which senses any spacecraft rotation about any of the three axes. Total attitude information goes to the CMC as well as to the FDAI's on the console. If a specific attitude or orientation is desired, attitude error signals are sent to the reaction jet engine control assembly. Then the proper reaction jet automatically fires in the direction necessary to return the spacecraft to the desired position.

The CMC provides primary control of thrust. The flight crew pre-sets thrusting and spacecraft data into the computer by means of the display keyboard. The forthcoming commands include time and duration of thrust. Accelerometers sense the amount of change in velocity obtained by the thrust. Thrust direction control is required because of center of gravity shifts caused by depletion of propellants in service propulsion tanks. This control is accomplished through electromechanical actuators which position the gimbaled SPS engine. Automatic control commands may originate in either the guidance and navigation subsystem or the SCS. There is also provision for manual controls.

Reaction Control Systems (RCS)

The Command Module and the Service Module each has its own independent system, the CM RCS and the SM RCS respectively. The SM RCS has four identical RCS "quads" mounted around the SM 90 degrees apart. Each quad has four 100-pound thrust engines, two fuel and two oxidizer tanks, and a helium pressurization sphere. The SM RCS provides redundant spacecraft attitude control through cross-coupling logic inputs from the Stabilization and Guidance Systems. Small velocity change maneuvers can also be made with the SM RCS. The CM RCS consists of two independent subsystems of six 94-pound thrust engines each. Both subsystems are activated after separation from the SM; one is used for spacecraft attitude control during entry. The other serves in standby as a backup. Propellants for both CM and SM RCS are monomethyl hydrazine fuel and nitrogen tetroxide oxidizer with helium pressurization. These propellants are hypergolic, i.e. they burn spontaneously when combined without need for an igniter.

Electrical Power System

The Electrical Power System (EPS) provides electrical energy sources, power generation and control, power conversion and conditioning, and power distribution to the spacecraft throughout the mission. The EPS also furnishes drinking water to the astronauts as a by-product of the fuel cells. The primary source of electrical power is the fuel cells mounted in the SM. Each cell consists of a hydrogen compartment, an oxygen compartment, and two electrodes. The cryogenic gas storage system, also located in the SM, supplies the hydrogen and oxygen used in the fuel cell power plants, as well as the oxygen used in the ECS.

Three silver-zinc oxide storage batteries supply power to the CM during entry and after landing, provide power for sequence controllers, and supplement the fuel cell's during periods of peak power demand. These batteries are located in the CM lower equipment bay. A battery charger is located in the same bay to assure a full charge prior to entry.

Two other silver-zinc oxide batteries, independent of and completely isolated from the rest of the dc power system, are used to supply power for explosive devices for CM/SM separation, parachute deployment and separation, third-stage separation, launch escape system tower separation, and other pyrotechnic uses.

Emergency Detection System

The Emergency Detection System (EDS) monitors critical conditions of launch vehicle powered flight. Emergency conditions are displayed to the crew on the main display console to indicate a necessity for abort. The system includes provisions for a crew-initiated abort with the use of the LES or with the SPS after tower jettison. The crew can initiate an abort separation from the LV from prior to lift-off until the planned separation time. A capability also exists for commanding early staging of the S-IVB from the S-II stage when necessary, Also included in the system are provisions for an automatic abort in case of the following time-critical conditions:

1. Loss of thrust on two or more engines on the first stage of the LV.

2. Excessive vehicle angular rates in any of the pitch, yaw, or roll planes.

3. Loss of "hotwire" continuity from SM to IU.

The EDS will automatically initiate an abort signal when two or more first-stage engines are out or when LV excessive rates are sensed by gyros in the IU. The abort signals are sent to the master events sequence controller, which initiates the abort sequence. The engine lights on the Main Display Console provide the following information to the crew: ignition, cutoff, engine below pre-specified thrust level, and physical stage separation. A yellow "S-II Sep" light will illuminate at second-stage first-plane separation and will extinguish at second plane separation. A high-intensity, red "ABORT" light is illuminated if an abort is requested by the Launch Control Center for a pad abort or an abort during liftoff via updata link. The "ABORT" light can also be illuminated after lift-off by the Range Safety Officer or by the Mission Control Center via the updata link from the Manned Space Flight Network.

Launch Escape System

General

The Launch Escape System (LES) (Figure 18) includes the LES structure, canards, rocket motors, and ordnance. The LES provides an immediate means of separating the CM from the LV during pad or suborbital aborts up through completion of second stage ignition. During an abort, the LES must provide a satisfactory earth return trajectory and CM orientation before jettisoning from the CM. The jettison or abort can be initiated manually or automatically.

Assembly

The forward or rocket section of the system is cylindrical and houses three solid propellant rocket motors and a ballast compartment topped by a nose cone and "O-ball" which measures attitude and flight dynamics of the space vehicle. The 500-pound tower is made of titanium tubes attached at the top to a structural skirt that covers the rocket exhaust nozzles and at the bottom to the CM by means of explosive bolts. A Boost Protective Cover (BPC) is attached to the tower and completely covers the CM. It has 12 "blowout" ports for the CM reaction control engines, vents, and an 8-inch window. This cover protects the CM from the rocket exhaust and also from the heat generated during launch vehicle boost. It remains attached to the tower and is carried away when the LES is jettisoned. Two canards mounted near the forward end of the assembly aerodynamically tumble the CM in the pitch plane during an abort so that the heat shield is forward.

Propulsion

Three solid propellant motors are used on the LES. They are:

1. The Launch Escape Motor which provides thrust for CM abort weighs 4700 pounds and provides 147,000 pounds of thrust at sea level for approximately eight seconds.

2. The Pitch Control Motor which provides an initial pitch maneuver toward the Atlantic Ocean during pad or low-altitude abort. It weighs 50 pounds and provides 2400 pounds of thrust for half a second.

3. The Tower Jettison Motor, which is used to jettison the LES, provides 31,500 pounds of thrust for one second.

System Operation

The system is activated automatically by the emergency detection system in the first 100 seconds or manually by the astronauts at any time from the pad to jettison altitude. The system is jettisoned at about 295,000 feet, or about 30 seconds after ignition of the second stage. After receiving an abort signal, the booster is cut off (after 30 seconds of flight), the CM-SM separation charges are fired, and the launch escape motor is ignited. The launch escape motor lifts the CM and the pitch control motor (used only at low altitudes) directs

LAUNCH ESCAPE SYSTEM

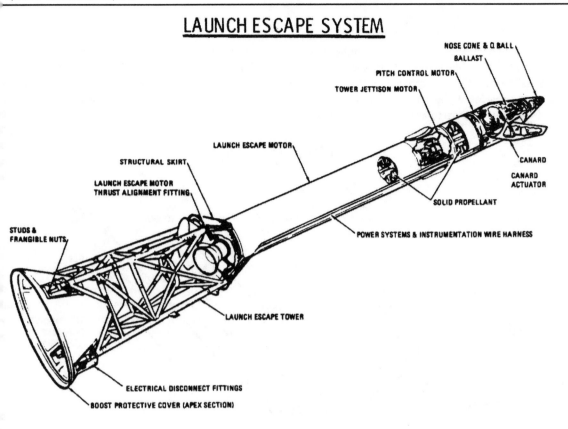

Fig. 18

the flight path off to the side. Two canards are deployed 11 seconds after an abort is initiated. Three seconds later on extreme low-altitude aborts, or at approximately 24,000 feet on high-altitude aborts, the tower separation devices are fired and the jettison motor is started. These actions carry the LES away from the CM's landing trajectory. Four-tenths of a second after tower jettisoning, the CM's earth landing system is activated and begins its sequence of operations to bring the CM down safely. All preceding automatic sequences can be prevented, interrupted, or replaced by crew action.

During a successful launch the LES is jettisoned by the astronauts, using the digital events timer and the "S-II Sep" light as cues. The jettisoning of the LES disables the Emergency Detection System automatic abort circuits. In the event of Tower Jettison Motor failure, the Launch Escape Motor may jettison the LES.

LUNAR MODULE

General

The Lunar Module (LM) (Figure 19) is designed to transport two men safely from the CSM, in lunar orbit, to the lunar surface and return them to the orbiting CSM. The LM provides operational capabilities such as communications, telemetry, environmental support, the transport of scientific equipment to the lunar surface, and the return of surface samples with the crew to the CSM. Physical characteristics are shown in Figure 20.

The Lunar Module consists of two stages; the Ascent Stage (AS), and the Descent Stage (DS). The stages are attached at four fittings by explosive bolts. Separable umbilicals and hard line connections provide subsystem continuity to operate both stages as a single unit until separate Ascent Stage operation is desired. The LM is designed to operate for 48 hours after separation from the CSM, with a maximum lunar stay time of 44 hours.

Ascent Stage

The Ascent Stage (AS) (Figure 21) accommodates two astronauts and is the control center of the LM. The stage structure provides three main sections consisting of a crew compartment and mid-section, which comprises the pressurized cabins and the unpressurized aft equipment bay. Other component parts of the structure consists of the Thrust Chamber Assembly (TCA) cluster supports, and antenna supports. The cylindrical crew compartment is of semi-monocoque, aluminum alloy construction. Large structural beams

LUNAR MODULE

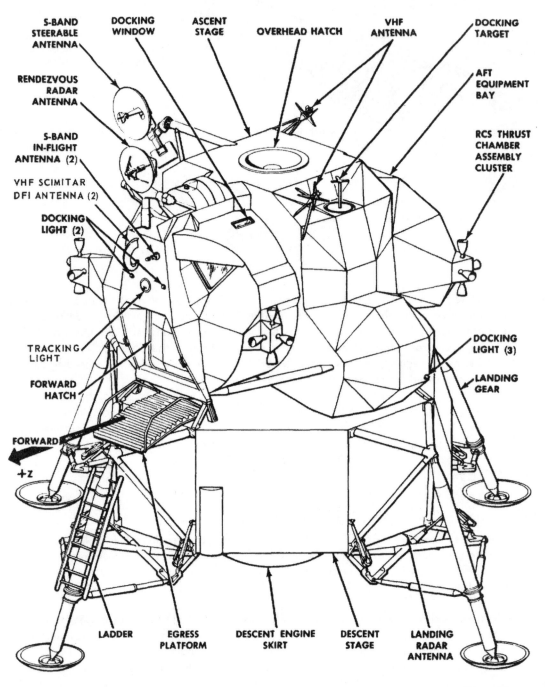

Fig. 19

LM PHYSICAL CHARACTERISTICS

TOTAL	
WEIGHT (Propellant & Crew)	32,025 LB.
WEIGHT (Less Prop)	9,336 LB.
ASCENT STAGE	
WEIGHT (Less Prop)	5,070 LB.
TANKED PROPELLANT (APS)	4,137 LB.
TANKED PROPELLANT (RCS)	608 LB.
DESCENT STAGE	
WEIGHT (Less Prop)	4,260 LB.
TANKED PROPELLANT (DPS)	17,944 LB.

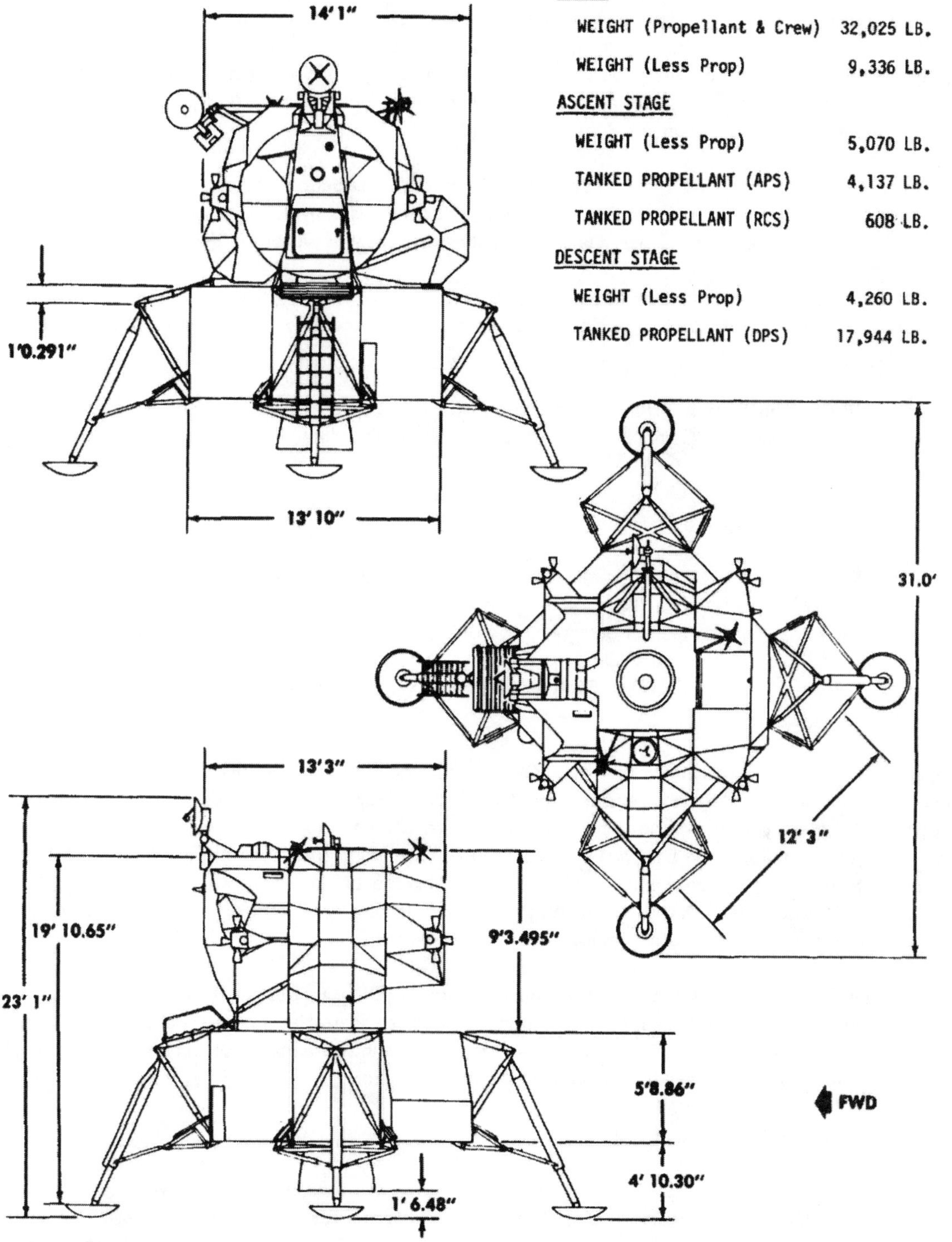

Fig. 20

extend up the front face and across the top of the crew compartment to distribute loads applied to the cabin structure. The structural concept utilizes beams, bulkheads, and trusses to "cradle" the cabin assembly. The cabin volume is approximately 235 cubic feet.

The entire Ascent Stage structure is enveloped by a vented blanket shield suspended at least two inches from the main structure. The thermal and micrometeoroid shield consists of multiple-layer aluminized Mylar, nickel foil, inconel mesh, inconel sheet, and, in certain areas, H-film. The shield nominally provides thermal insulation against +350° F temperatures; with H-film, protection up to +1000°F is provided.

The flight station area has two front windows, a docking window, window shades, supports and restraints, an Alignment Optical Telescope (AOT), Crewman Optical Alignment Sight (COAS), data files, and control and display panels. Two hatches are provided for ingress and egress. The inward-opening forward hatch is used for EVA exit and entry. The overhead hatch seals the docking tunnel which is used for the transfer of crew and equipment internally between the docked CSM and LM.

The Ascent Stage is the nucleus of all LM systems. Two Portable Life Support Systems are stowed in the LM and provisions have been made for their replenishment. Stowage is provided for docking equipment, Extra Vehicular Visors, Extra Vehicular Gloves, Lunar Overshoes, and crew provisions in general.

The Ascent Stage also provides external mounting for a CSM-active docking target, tracking and orientation lights, two VHF antennas, two S-band in-flight antennas, an S-band steerable antenna and a rendezvous radar antenna.

The Ascent Propulsion System (APS) provides for major +X axis translations when separated from the descent stage and a Reaction Control System (RCS) provides attitude and translational control about and along three axes.

Descent Stage

The Descent Stage (DS) (Figure 22) is the unmanned portion of the LM. It provides for major velocity changes of the LM to deorbit and land on the lunar surface. The basic structure consists of four main crossed-beams whose ends define the octagon shape of the stage. The major structural material is aluminum alloy. Thermal and micrometeoroid shielding is similar to that used on the Ascent Stage but with additional base heat shielding of nickel foil, H-film, Fibrocel, and Fiberfrox protecting the stage base from engine heat radiation.

The Descent Stage has four landing gear to absorb landing shock and to support the Descent Stage which must serve as a launch pad for the Ascent Stage. The Descent Stage engine nozzle extension is designed to collapse up to 28 inches and will not have any influence on LM lunar surface stability. Impact attenuation is achieved by compression of the main struts against crushable aluminum honeycomb. The landing gear trusses also provide the structural attachment points for securing the LM to the lower (fixed) portion of the Spacecraft LM Adapter (SLA). A ladder, integral to a primary landing gear start, provides access from and to the lunar surface from the ten-foot high forward hatch platform.

The Descent Stage contains the Descent Propulsion System (DPS) as well as electrical batteries, landing radar, supplements for the Environmental Control System, six batteries for the Portable Life Support Systems, a storage area for scientific equipment, an erectable S-band antenna, pyrotechnics, and generally, LM components not required for the lunar ascent phase of the mission.

Guidance, Navigation, and Control System

The Guidance, Navigation, and Control System (GN&CS) provides vehicle guidance, navigation, and control required for a manned lunar landing mission. The GN&CS utilizes a Rendezvous Radar (RR), and a Landing Radar (LR) to aid in navigation. The major subsystems of the GN&CS system are designated Primary Guidance and Navigation Subsystem (PGNS), Abort Guidance Subsystem (AGS), and the Control Electronics Subsystem (CES).

The GN&CS has a primary and alternate system path. The primary guidance path comprises the Primary Guidance and Navigation Subsystem, Control Electronics Subsystem, Landing Radar, Rendezvous Radar, and the selected propulsion system. The alternate system path comprises the Abort Guidance Subsystem, Control Electronics Subsystem, and the selected propulsion system. The term Primary Guidance, Navigation, and

LM ASCENT STAGE

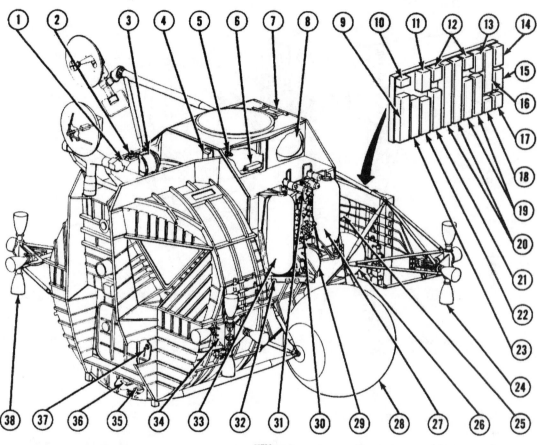

KEY

1. Abort sensor assembly
2. Alignment optical telescope
3. Inertial measurement unit
4. Pulse torque assembly
5. Cabin dump and relief valve (upper hatch)
6. CSM/LM electrical umbilical fairing
7. Aft equipment bay bulkhead
8. Water tank
9. Rendezvous radar electronics assembly
10. Propellant quantity gaging system control unit
11. Caution and warning electronics assembly
12. Electrical control assembly
13. Attitude and translation control assembly
14. S-band power amplifier and diplexer
15. S-band transceiver
16. Abort electronic assembly
17. Signal processor assembly
18. VHF transceiver and diplexer
19. Inverter
20. Batteries

21. Signal-conditioning and electronic replaceable assembly No. 2
22. Pulse-code-modulation and timing equipment assembly
23. Signal-conditioning and electronic replaceable assembly No. 1
24. RCS quadrant 2
25. Gaseous oxygen tank
26. Helium tank
27. RCS fuel tank
28. APS fuel tank
29. RCS helium tank
30. RCS tank module
31. Helium pressurization module
32. Oxidizer service panel
33. RCS oxidizer tank
34. RCS quadrant 1
35. Lighting control assembly
36. Auxiliary switching relay box
37. Cabin dump and relief valve (forward hatch)
38. RCS quadrant 4

Fig. 21

Control System (PGNCS) appears in certain technical mission documentation and connotes use of systems in the primary path of the LM Guidance, Navigation, and Control System.

LM DESCENT STAGE

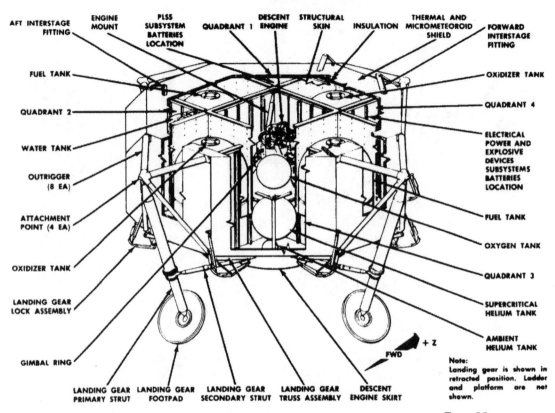

Fig. 22

Primary Guidance and Navigation Subsystem

The Primary Guidance and Navigation Subsystem (PGNS) establishes an inertial reference for guidance with an Inertial Measurement Unit, uses optics and radar for navigation, and a digital LM Guidance Computer (LGC) for data processing and generation of flight control signals. The inertially stabilized accelerometers sense incremental changes of velocity and attitude. Comparison of sensed instantaneous conditions against software programs generates corrections used to control the vehicle. The reference for the inertial system is aligned using the Alignment Optical Telescope, stars, horizons, and the computer. The PGNS, in conjunction with the CES, controls LM attitude, ascent or descent engine firing, descent engine thrust, and thrust vector. Control under the PGNS mode ranges from fully automatic to manual.

Abort Guidance Subsystem

The Abort Guidance Subsystem (AGS) provides an independent backup for the PGNS. The section is not utilized during aborts unless the PGNS has failed. The AGS is capable of determining trajectories required for a co-elliptic rendezvous sequence to automatically place the vehicle in a safe parking/rendezvous orbit with the CSM or can display conditions to be acted upon by the astronauts to accomplish rendezvous. The activated AGS performs LM navigation, guidance, and control in conjunction with the Control Electronics Subsystem (CES). The AGS differs from the PGNS in that its inertial sensors are rigidly mounted with respect to the vehicle rather than on a stabilized platform. In this mode, the Abort Sensor Assembly (ASA) measures attitude and acceleration and supplies data to the Abort Electronics Assembly (AEA) which is a high-speed digital computer.

Control Electronics Subsystem

The Control Electronics Subsystem (CES) controls LM attitude and translation about and along three axes by processing commands from the PGNS or AGS and routing on/off commands to 16 reaction control engines, ascent engine, or descent engine. Descent engine thrust vector is also controlled by the CES.

Rendezvous Radar

The Rendezvous Radar (RR) tracks the CSM to provide relative line of sight, range and range rate data for rendezvous and docking. The transponder in the CSM augments the transmitted energy of the RR thus increasing radar capabilities and minimizing power requirements. Radar data is automatically entered into the LGC in the PGNS mode. During AGS operation, data inputs are entered into the Abort Electronics Assembly (AEA) through the Data Entry and Display Assembly (DEDA) by the crew from cabin displays. Radar data is telemetered to the Manned Space Flight Network and monitored for gross inaccuracies.

Landing Radar

The Landing Radar provides the LGC with slant range and velocity data for control of the descent to the lunar surface. Slant range data is available below lunar altitudes of approximately 25,000 feet and velocity below approximately 18,000 feet.

Main Propulsion

Main Propulsion is provided by the Descent Propulsion System (DPS) and the Ascent Propulsion System (APS). Each system is wholly independent of the other. The DPS provides the thrust to control descent to the lunar surface. The APS can provide the thrust for ascent from the lunar surface. In case of mission abort, the APS and/or DPS can place the LM into a rendezvous trajectory with the CSM from any point in the descent trajectory. The choice of engine to be used depends on the cause for abort, on how long the descent engine has been operating, and on the quantity of propellant remaining in the Descent Stage. Both propulsion systems use identical hypergolic propellants. The fuel is a 50-50 mixture of Hydrazine and Unsymmetrical Dimethyl Hydrazine and the oxidizer is Nitrogen Tetroxide. Gaseous Helium pressurizes the propellant feed systems. Helium storage in the DPS is at cryogenic temperatures in the super-critical state and in the APS it is gaseous at ambient temperatures.

Ullage for propellant settling is required prior to descent engine start and is provided by the +X axis reaction engines. The descent engine is gimbaled, throttleable and restartable. The engine can be throttled from 1050 pounds of thrust to 6300 pounds. Throttle positions above this value automatically produce full thrust to reduce combustion chamber erosion. Nominal full thrust is 9870 pounds. Gimbal trim of the engine compensates for a changing center of gravity of the vehicle and is automatically accomplished by either the PGNS or AGS. Automatic throttle and on/off control is available in the PGNS mode of operation. The AGS commands on/off operation but has no automatic throttle control capability. Manual control capability of engine firing functions has been provided. Manual thrust control override may, at any time, command more thrust than the level commanded by the LGC.

The ascent engine is a fixed, non-throttleable engine. The engine develops 3500 pounds of thrust, sufficient to abort the lunar descent or to launch the Ascent Stage from the lunar surface and place it in the desired lunar orbit. Control modes are similar to those described for the descent engine. The Ascent Propulsion System propellant is contained in two spherical titanium tanks, one for oxidizer and the other for fuel. Each tank has a volume of 36 cubic feet. Total fuel weight is 2008 pounds of which 71 pounds are unusable. Oxidizer weight is 3170 pounds of which 92 pounds are unusable. The APS has a limit of 35 starts, must have a propellant bulk temperature between 50°F and 90°F prior to start, must not exceed 460 seconds of burn time and has a system life of 24 hours after pressurization.

In general, the main propulsion systems use pyrotechnic isolation valves in pressurization and propellant lines to prevent corrosive deterioration of components. Once the APS or DPS is activated, its reliable operating time is limited but adequate for its designed use.

Reaction Control System

The Reaction Control System (RCS) stabilizes the LM, provides ullage thrust for the DPS or APS, helps to maintain the desired trajectory during descent, and controls LM attitude and translation about or along three axes during hover. Sixteen engines termed Thrust Chamber Assemblies (TCA's) of 100 pounds thrust each are mounted symmetrically around the LM Ascent Stage in clusters of four. The RCS contains two independent, parallel systems (A&B) controlling two TCA's in each cluster. Each system, operating alone, can perform all required attitude control requirements, however translational performance is slightly degraded under single system operation. The independent propellant systems have a cross feed capability for increased operational dependability. During APS thrusting, APS propellant can supplement the RCS system. The propellant tanks utilize bladders to achieve positive expulsion feed under zero-g gravity conditions. Malfunctioning TCA pairs can be deactivated by manual switches.

The RCS TCA firing is accomplished by the Control Electronics Section of the GN&CS in response to manual commands or signals generated in the PGNS or AGS modes. RCS modes of operation are: automatic; attitude hold (semi-automatic); and manual override. The TCA's firing time ranges from a pulse of less than one second up to steady state operation.

Thirty-two heaters are used to heat the 16 TCA's. TCA temperature requirements ranging from 132°F to 154°F are important to safe and proper TCA operation. Propellant capacity of each system of the RCS is: oxidizer (N_2O_4) 207.5 pounds, 194.9 pounds usable; Fuel (50-50 N_2H_4 and UDMH) 106.5 pounds, 99.1 pounds usable.

In order to ensure reliable RCS operation, firing time for each TCA must not exceed 500 seconds with firing times exceeding one second, and 1000 seconds of pulses with firing times less than one second. RCS operation requires propellant tank temperatures between 40°F and 100°F. Firing time of vertically mounted thrusters is limited to prevent damage to descent stage insulation or the ascent stage antennas.

Electrical Power System

The Electrical Power System (EPS) contains six batteries which supply the electrical power requirements of the LM during undocked mission phases. Four batteries are located in the Descent Stage (DS) and two in the Ascent Stage (AS). Batteries for the Explosive Devices System are not included in this system description. Post launch LM power is supplied by the IDS batteries until the LM and CSM are docked. While docked, the CSM supplies electrical power to the LM up to 296 watts (Peak). During the lunar descent phase, the two AS batteries are paralleled with the DS batteries for additional power assurance. The DS batteries are utilized for LM lunar surface operations and checkout. The AS batteries are brought on the line just before ascent phase staging. All batteries and busses may be individually monitored for load, voltage, and failure. Several isolation and combination modes are provided.

Two Inverters, each capable of supplying full load, convert the dc to ac for 115-volt, 400-hertz supply. Electrical power is distributed by the following buses: LM Pilot's dc bus, Commanders dc bus, and ac buses A&B.

The four Descent Stage silver-zinc oxide batteries are identical and have a 400 ampere-hour capacity at 28 volts. Because the batteries do not have a constant voltage at various states of charge/load levels, "high" and "low" voltage tops are provided for selection. The "low voltage" tap is selected to initiate use of a fully charged battery. Cross-tie circuits in the busses facilitate an even discharge of the batteries regardless of distribution combinations. The two silver zinc oxide Ascent Stage batteries are identical to each other and have a 296 ampere-hour capacity at 28 volts. The AS batteries are normally connected in parallel for even discharge. Because of design load characteristics, the AS batteries do not have and do not require high and low voltage taps.

Nominal voltage for AS and DS batteries is 30.0 volts. Reverse current relays for battery failure are one of many components designed into the EPS to enhance EPS reliability. Cooling of the batteries is provided by the Environmental Control System cold rail heat sinks. Available ascent electrical energy is 17.8 kilowatt hours at a maximum drain of 50 amps per battery and descent energy is 46.9 kilowatt hours at a maximum drain of 25 amps per battery.

Environmental Control System

The Environmental Control System (ECS) provides a habitable environment for two astronauts for a maximum of 48 hours while the LM is separated from the CSM. Included in this capability is four cabin decompression/ re-pressurization cycles. The ECS also controls the temperature of electrical and electronic equipment, stores and provides water for drinking, cooling, fire extinguishing, and food preparation. Two oxygen and two water tanks are located in the Ascent Stage. One larger oxygen tank and a larger water tank is located in the Descent Stage.

The ECS is comprised of an Atmosphere Revitalization Subsystem (ARS), an Oxygen Supply and Cabin Pressure Control Subsystem (OSCPCS), a Water Management Subsystem (WMS), a Heat Transport Subsystem (HTS), and an oxygen and water supply to the Portable Life Support System (PLSS). The ARS cools and ventilates the Pressure Garment Assemblies, controls oxygen temperature, and the level of carbon dioxide in the atmosphere, removes odors, particles, noxious gases, and excess moisture.

Oxygen Supply and Cabin Pressure Control Subsystem

The Oxygen Supply and Cabin Pressure Control Subsystem (OSCPCS) stores gaseous oxygen and maintains cabin and suit pressure by supplying oxygen to the ARS. This replenishes losses due to crew metabolic consumption and cabin or suit leakage. The oxygen tank in the Descent Stage provides oxygen during the descent and lunar-stay phases of the mission, and the two in the ascent stage are used during the ascent and rendezvous phases of the mission.

Water Management Subsystem

The Water Management Subsystem (WMS) supplies water for drinking, cooling, fire extinguishing, and food preparation; for refilling the PLSS cooling water tank; and for pressurization of the secondary coolant loop of the HTS. It also provides for delivery of water from ARS water separators to HTS sublimators; and from the water tanks to ARS and HTS sublimators. The water tanks are pressurized before launch to maintain the required pumping pressure in the tanks. The Descent Stage tank supplies most of the water required until staging occurs. After staging, water is supplied by the two Ascent Stage storage tanks. A self-sealing "PLSS DRINK" valve delivers water for drinking and food preparation.

Heat Transport Subsystem

The Heat Transport Subsystem (HTS) consists of a primary coolant loop and a secondary coolant loop. The secondary loop serves as a backup loop and functions in the event the primary loop fails. A water-glycol solution circulates through each loop. The primary loop provides temperature control for batteries, electronic equipment that require active thermal control, and for the oxygen that circulates through the cabin and pressure suits. The batteries and electronic equipment are mounted on cold plates and rails through which coolant is routed to remove excess heat.

The cold plates used for equipment required for mission abort contain two separate coolant passages, one for the primary loop and one for the secondary loop. The secondary coolant loop serves only abort equipment cold plates.

In flight, excess heat rejection from both coolant loops is achieved by the primary and secondary sublimators which are vented overboard. A coolant pump recirculation assembly contains all the HTS coolant pumps and associated check and relief valves. Coolant flow from the assembly is directed through parallel circuits to the cold plates for the electronic equipment and the oxygen to glycol heat exchanger in the ARS.

Communications System

The Communications System (CS) provides the links between the LM and the Manned Space Flight Network (MSFN), between the LM and the CSM, and between the LM and any extravehicular astronaut. The following information is handled by the Communications System: Tracking and ranging; voice; PCM telemetry (LM status); biomedical data; computer updates; morse code; television; EVA/LM EMU data; and LM/CSM telemetry retransmission. The communications links and their functions are listed in Figure 23. The CS includes all S-band, VHF, and signal processing equipment necessary to transmit and receive voice, tracking, and ranging data, and to transmit telemetry and emergency keying.

Fig. 23 LM COMMUNICATIONS LINKS

Link	Mode	Band	Purpose
MSFN-LM-MSFN	Pseudo random noise	S-band	Ranging and tracking
LM-MSFN	Voice	S-band, 'VHF (optional)	In-flight communications
LM-CSM	Voice	VHF simplex	In-flight communications
CSM-LM-MSFN	Voice	VHF and S-band	Conference (with LM as relay)
LM-CSM	Low-bit-rate telemetry	VHF (one way)	Record and retransmit to earth
MSFN-LM	Voice	S-band, VHF (optional)	In-flight communications
MSFN-LM	Uplink data or uplink voice backup	S-band	Update LGC or voice backup for in-flight communications
LM-MSFN	Biomed- PCM telemetry	S-band	In-flight communications
LM-MSFN-CSM	Voice	S-band or VHF	Conference (with earth as relay)
LM-EVA-LM	Voice and data	VHF duplex	EVA direct communication
EVA-LM-MSFN	Voice and data	VHF, S-band	Conference (with LM as relay)
EVA-LM-CSM	Voice and data	VHF duplex	Conference (with LM as relay)
CSM-MSFN- LM-EVA	Voice and data	S-band, VHF	Conference (via MSFN- relay)

The CS antenna equipment consists of: two S-band in-flight antennas; an S-band steerable antenna; two VHF in-flight antennas and Diplexer, and RF selector switches for S-band and VHF. The "line of sight" range of the VHF transmitter is limited to 740 nautical miles. The LM S-band capability covers earth-lunar distances.

Explosive Devices System

The Explosive Devices System (EDS) uses explosives to activate or enable various LM equipment. The system deploys the landing gear, enables pressurization of the descent, ascent, and RCS propellant tanks, venting of descent propellant tanks, and separation of the Ascent and Descent Stages. There are two separate systems in the EDS. The systems are parallel and provide completely redundant circuitry. Each system has a 37.1 volt (no load) battery, relays, time delay circuits, fuse resistors, buses and explosive cartridges.

Two separate cartridges are provided for each EDS function. Each cartridge is sufficient to perform the function without the other. The EDS supports the main propulsion systems by clearing the valves isolating pressurants and propellants. Other pyrotechnic devices guillotine interstage umbilicals in addition to the structural connections. System performance is indicated to the crew by instrumentation and to the MSFN by telemetry. The two EDS batteries use silver-zinc plates and are rated at 75 ampere-hour. Battery output/voltage status is displayed to the crew. One battery is located in the Descent Stage and one is in the Ascent Stage.

Instrumentation System

The Instrumentation System (IS) monitors the LM subsystems, performs in-flight checkout, prepares LM status data for transmission to the MSFN, provides timing frequencies and correlated data for LM subsystems, and stores voice and time correlation data. During the lunar mission, the IS performs lunar surface LM checkout and provides scientific instrumentation for lunar experiments.

The IS consists of system sensors, a Signal Conditioning Electronics Assembly (SCEA),

Pulse-Code-Modulation and Timing Electronics Assembly (PCMTEA), Caution and Warning Electronics Assembly (CWEA), and a Data Storage Electronics Assembly (DSEA). The CWEA provides the astronauts and MSFN with a continuous rapid check of data supplied by the SCEA for malfunction detection. The CWEA provides signals that light caution lights, warning lights, component caution lights, and "Master-Alarm" pushbutton lights.

Lighting

Interior lighting is designed to enhance crew performance by reducing crew fatigue in an environment of interior-exterior glare effects. Exterior lighting includes a radioluminescent docking target, five docking lights, and a high intensity tracking light. The five docking lights are automatically turned on prior to the first CSM docking and are turned off after docking. They indicate gross relative attitude of the vehicle and are color discernable to a distance of 1000 feet. The flashing, high-intensity, tracking light on the LM facilitates CSM tracking of the LM. It has a beam spread of 60 degrees and flashes 60 times per minute.

Crew Provisions

Apparel

The combination of items a crewman wears varies during a mission (Figure 24). There are three basic configurations of dress: unsuited, suited, and extravehicular. A brief description of each item is contained in the latter part of this section.

Unsuited

This mode of dress is worn by crewmen in the CSM under conditions termed "shirt-sleeve environment." The crewman wears a biomedical harness, a Communications Carrier, a Constant Wear Garment, Flight Coveralls, and Booties. This unsuited mode is the most comfortable, convenient, and consequently, the least fatiguing of the modes. When unsuited, the astronaut relies upon the CSM ECS to maintain the proper cabin environment of pressure, temperature, and oxygen.

Suited

This mode enables a crewman to operate in an unpressurized cabin up to the design life of the pressure suit of 115 hours. The intravehicular configuration includes: The Pressure Garment Assembly (PGA) made up of a Torso-Limb Suit, Pressure Helmet, and Pressure Gloves; the Fecal Containment System; Constant Wear Garment; Biomedical Belt; Communications Carrier; Urine Collection and Transfer Assembly, and a PGA integrated with a thermal micrometeoroid garment.

The Command Module Pilot does not participate in any extravehicular activity, permitting substitution of a lighter, fire-resistant covering over the PGA in lieu of the thermal-micrometeoroid garment. Various suit fittings and hardware required for LM and EVA operations are also omitted from the Command Module Pilot's suit.

Extravehicular

In the extravehicular configuration, the Constant Wear Garment is replaced by a Liquid Cooling Garment and four items are added to the Pressure Garment Assembly: Extravehicular Visor Assembly, Extravehicular Gloves, Lunar Overshoes, and a connector cover which fits over umbilical connections on the front of the suit. The addition of the Portable Life Support System (PLSS) and Oxygen Purge System back-pack completes the configuration termed the Extravehicular Mobility Unit (EMU). The EMU protects the astronaut from radiation, micrometeorite impact, and lunar surface temperatures ranging from +250°F to -250°F.

APOLLO APPAREL

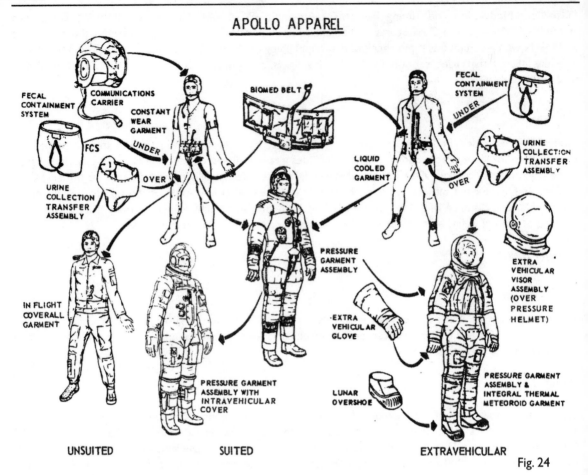

Fig. 24

Item Description

Torso Limb Suit Assembly The Torso Limb Suit is the basic pressure envelope for the astronaut. It contains connectors for oxygen, water (for the Liquid Cooling Garment), communication, biomedical data, and urine transfer.

Pressure Helmet - The Pressure Helmet is basically a polycarbonate plastic shell. It contains a vent manifold and an air-tight feed port for eating, drinking, and purging. The astronaut can turn his head within the fixed helmet.

Pressure Glove - The Pressure Glove is basically made of nylon tricot dipped in Neoprene. A fingerless glove, inner and outer covers, and a restraint system complete the assembly. The Extravehicular Glove is a modified pressure glove with additional layers of thermal and protective material added.

Integrated Thermal Meteoroid Garment - This garment is sewn over the Torso Limb Suit. Construction utilizes multi-layered combinations of Beta cloth, aluminized Kaplan film, Beta Marquisette, Neoprene-coated nylon Ripstop, and Chromel-R. Snap-secured covers are located for inner access to some PGA areas and pockets are provided for specified items. LM restraint rings are integrated into the hip area. Boots are attached over the PGA with slide fasteners and loop tape.

Lunar Overshoe - The overshoe is worn over the PGA thermal, meteoroid covered boot. The Lunar Overshoe meets the extensive, additional, thermal and protective requirement for a lunar excursion. Materials used in its construction are: teflon-coated Beta cloth, Kapton film Beta Marquisette, Beta felt, silicon rubber and Chromel-R.

Extravehicular Visor - The Extravehicular Visor consists of two pivoted polycarbonate visors mounted on a polycarbonate shell. The visors furnish protection against micrometeoroids, solar heat and radiation, and protection of the PGA helmet. The outer visor features a vacuum-deposited gold film.

Liquid Cooling Garment - The Liquid Cooling Garment consists of a network of Tygon tubing interwoven in nylon Spandex material. Water from the PLSS circulates through the tubing to maintain the desired suit temperature. An inner liner is fabricated from nylon chiffon. The integral socks do not contain cooling tubes.

Constant Wear Garment - The Constant Wear Garment is an undergarment for the flight coveralls and the non-EVA spacesuit configuration. It is fabricated in one piece, encloses the feet, has short sleeves, a waist to neck zipper, and lower torso openings front and rear.

Flight Coverall - The flight coverall is the outer garment for unsuited operation. It is of two-piece, Beta cloth construction with zipper and pockets.

Booties - Booties worn with the flight coveralls are made of Beta cloth, with Velcro hook material bonded to the soles. During weightlessness, the Velcro hook engages Velcro pile patches attached to the floor to hold the crewman in place.

Communications Carrier and Biomedical Harness - The Communications Carrier is a polyurethane foam headpiece which positions two independent earphones and microphones. The Biomedical Harness carries signal conditioners and converters to transmit heart beat and respiration rates of the astronauts. The wiring of the Biomedical Harness and Communications Carrier connect to a common electrical connector which interfaces with the PGA or an adapter when unsuited.

Urine Collection and Transfer Assembly - The Urine Collection and Transfer Assembly is a truss-like garment which functions by use of a urinal cuff, storage compartment, and tube which connects to the external collection system. It is worn over the Constant Wear Garment or Liquid Cooling Garment.

Fecal Containment System - The Fecal Containment System (FCS) is an elastic underwear with an absorbent liner around the buttock area. This system is worn under the LCG or CWG to allow emergency defecation when the PGA is pressurized. Protective ointment is used on the buttocks and perineal area to lessen skin irritation.

Portable Life Support System

The Portable Life Support System (PLSS) is a portable, self-powered, rechargeable environmental control system with a communications capability. It is carried as a backpack in the extravehicular suited mode. It weighs about 68 pounds. The PLSS supplies pressurized oxygen to the PGA, cleans and cools the suit atmosphere, cools and circulates water through the Liquid Cooling Garment, and provides RF communications with a dual VHF transceiver. The PLSS can operate for up to four hours in a space environment before replenishment of water and oxygen is required. The 17-volt PLSS battery can supply 280 watt-hours of electrical power to meet a nominal usage rate of 50 watts per hour.

Oxygen Purge System

A detachable, non-rechargeable oxygen purge system attaches to the top of the PLSS. The system can supply 30 minutes of regulated flow to the PGA independent of the PLSS for contingency operations. The Oxygen Purge System may be removed from the PLSS and used as an emergency source of oxygen at any time. The Oxygen Purge System also serves as a mount for the PLSS antenna.

Food and Water

Food supplies in the LM and CSM are designed to supply each astronaut with a balanced diet of approximately 2800 calories per day. The food is either freeze dried or concentrated and is carried in vacuum-packaged plastic bags. Each bag of freeze-dried food has a one-way valve through which water is inserted and a second valve through which food passes. Concentrated food is packaged in bite size units and needs no reconstitution. Several bags are packaged together to make one meal bag. The meal bags have red, white, and blue dots to identify them for each crewman, as well as labels to identify them by day and meal.

The food is reconstituted by adding hot or cold water through the one-way valve. The astronaut kneads the bag and then cuts the neck of the bag and squeezes the food into his mouth. A "Feed Port" in the Pressure Helmet allows partaking of liquid food and water while suited. Food preparation water is dispensed from a unit which supplies 150°F and 50°F water in the CSM and 90°F and 50°F water in the LM.

Drinking water comes from the water chiller to two outlets, the water meter dispenser, and the food preparation unit. The dispenser has an aluminum mounting bracket, a 72-inch coiled hose, and a dispensing valve unit in the form of a button actuated pistol. The pistol barrel is placed in the mouth and the button is pushed for each half-ounce of water. The meter records the amount of water drunk. A valve is provided to shut off the system in case the dispenser develops a leak or malfunction.

Couches and Restraints

Command/Service Module

The astronaut couches are individually adjustable units made of hollow steel tubing and covered with a heavy, fireproof, fiberglass cloth. The couches rest on a head beam and two side-stabilizer beams supported by eight attenuator struts (two each for the Y and Z axes and four for the X axis) which absorb the impact of landing. These couches support the crewmen during acceleration and deceleration, position the crewmen at their duty stations, and provide support for translation and rotation hand controls, lights, and other equipment.

The couches can be folded or adjusted into a number of seat positions. The one used most is the 85-degree position assumed for launch, orbit entry, and landing. The 170-degree (flat-out) position is used primarily for the center couch, so that crewmen can move into the lower equipment bay. The armrests on either side of the center couch can be folded footward so the astronauts from the two outside couches can slide over easily. The hip pan of the center couch can be disconnected and the couch can be pivoted around the head beam and laid on the aft bulkhead floor of the CM. This provides both room for the astronauts to stand and easier access to the side hatch for extravehicular activity.

Two armrests are attached to the back pan of the left couch and two armrests are attached to the right couch. The center couch has no armrests. The translation and rotation controls can be mounted to any of the four armrests. A support at the end of each armrest rotates 100 degrees to provide proper tilt for the controls. The couch seat pan and leg pan are formed of framing and cloth, and the foot pan is all steel. The foot pan contains a restraint device which holds the foot in place.

The couch restraint harness consists of a lap belt and two shoulder straps which connect to the lap belt at the buckle. The shoulder straps connect to the shoulder beam of the couch. Other restraints in the CM include handholds, a hand bar, hand straps, and patches of Velcro which hold the crewmen when they wear booties.

The astronauts may sleep in bags under the left and right couches with heads toward the hatch or in their couches. The two sleeping bags are made of lightweight Beta fabric 64 inches long, with zipper openings for the torso and a 7-inch diameter opening for the neck. They are supported by two longitudinal straps that attach to storage boxes in the lower equipment bay and to the CM inner structure. The astronauts sleep in the bags when unsuited and restrained on top of the bags when suited.

Lunar Module

The crew support and restraint equipment in the LM includes armrests, hand holds, Velcro on the floor to interface with the PGA Boots, and a restraint assembly operated by a rope-and-pulley arrangement that holds the LM crewmen in a standing position. The restraint assembly attaches to "D" rings located at the hips of the astronaut's suit and holds him to the cabin floor with a force of about 30 pounds (Figure 25). The armrests restrain the crewmen laterally. LM crew members rest positions are shown in Figure 26.

LM CREWMAN AT FLIGHT STATION

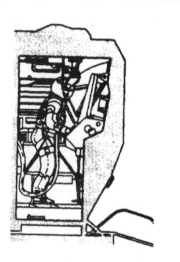

Fig. 25

LM CREWMEN REST POSITIONS

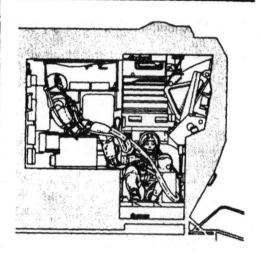

Fig. 26

Hygiene Equipment

Hygiene equipment includes wet and dry cloths for cleaning, towels, a toothbrush, and the waste management system. The waste management system controls and disposes of waste solids, liquids, and gases. The major portion of the system is in the right-hand equipment bay. The system stores feces, removes odors, dumps urine overboard, and removes urine from the space suit. Waste management in the LM differs in that urine is stored and not dumped overboard.

Operational Aids

Operational aids include data files, tools, workshelf, cameras, fire extinguishers, oxygen masks, medical supplies, and waste bags. The CM has one fire extinguisher, located adjacent to the left-hand and lower equipment bays. The extinguisher weighs about eight pounds. The extinguishing agent is an aqueous gel expelled in two cubic feet of foam for approximately 30 seconds at high pressure. Fire ports are located at various panels so that the extinguisher's nozzle can be inserted to put out a fire behind the panel.

Oxygen masks are provided for each astronaut in case of smoke, toxic gas, or other hostile atmosphere in the cabin while the astronauts are out of their suits in the CM. Oxygen is supplied through a flexible hose from the emergency oxygen/repressurization unit in the upper equipment bay.

Medical supplies are contained in an emergency medical kit, about 7 x 5 x 5 inches, which is stored in the lower equipment bay. It contains oral drugs and pills (pain capsules, stimulant, antibiotic, motion sickness, diarrhea, decongestant, and aspirin), injectable drugs (for pain and motion sickness), bandages, topical agents (first-aid cream, sun cream, and an antibiotic ointment), and eye drops.

Survival Equipment

Survival equipment, intended for use in an emergency after earth landing, is stowed in two rucksacks in the right-hand forward equipment bay. One of the rucksacks contains a three-man rubber life raft with an inflation assembly, a carbon-dioxide cylinder, a sea anchor, dye marker, and a sunbonnet for each crewman. The other rucksack contains a beacon transceiver, survival lights, desalter kits, a machete, sun glasses, water cans, and a medical kit. The survival medical kit contains the same type of supplies as the emergency medical kit: six bandages, six injectors, 30 tablets, and one tube of all-purpose ointment.

Miscellaneous Equipment

Each crewman is provided a toothbrush, wet and dry cleansing cloths, ingestible toothpaste, a 64-cubic inch container for personal items, and a two-compartment temporary storage bag. A special tool kit is provided which also contains three jack screws for contingency hatch closure.

LAUNCH COMPLEX

GENERAL

Launch Complex 39 (LC 39), located at Kennedy Space Center, Florida, is the facility provided for the assembly, checkout, and launch of the Apollo Saturn V Space Vehicle. Assembly and checkout of the vehicle is accomplished on a Mobile Launcher in the controlled environment of the Vehicle Assembly Building. The Space Vehicle and the Mobile Launcher are then moved as a unit by the Crawler-Transporter to the launch site. The major elements of the launch complex shown in Figure 27 are the Vehicle Assembly Building (VAB), the Launch Control Center (LCC), the Mobile Launcher (ML), the Crawler-Transporter (C/T) the crawlerway, the Mobile Service Structure (MSS), and the launch pad.

LC 39 FACILITIES AND EQUIPMENT

Vehicle Assembly Building

The VAB provides a protected environment for receipt and checkout of the propulsion stages and IU, erection of the vehicle stages and spacecraft in a vertical position on the ML, and integrated checkout of the assembled space vehicle. The VAB, as shown in Figure 28, is a totally-enclosed structure covering eight acres of ground. It is a structural steel building approximately 525 feet high, 518 feet wide, and 716 feet long. The principal operational elements of the VAB are the low bay and high bay areas. A 92-foot wide transfer aisle extends through the length of the VAB and divides the low and high bay areas into equal segments. The low bay area provides the facilities for receiving, uncrating, checkout, and preparation of the S-II stage, S-IVB stage, and the IU. The high bay area provides the facilities for erection and checkout of the S-IC stage; mating and erection operations of the S-II stage, S-IVB stage, IU, and Spacecraft; and integrated checkout of the assembled Space Vehicle. The high bay area contains four checkout bays, each capable of accommodating a fully-assembled Apollo Saturn V Space Vehicle.

Launch Control Center

The LCC, Figure 28, serves as the focal point for overall direction, control, and monitoring of space vehicle checkout and launch. The LCC is located adjacent to the VAB and at a sufficient distance from the launch pad (three miles) to permit the safe viewing of lift-off without requiring site hardening.

The LCC is a four-story structure. The ground floor is devoted to service and support functions. The second floor houses telemetry and tracking equipment, in addition to instrumentation and data reduction facilities. The third floor is divided into four separate but similar control areas, each containing a firing room, a computer room, a mission control room, a test conductor platform area, a visitor gallery, and offices. The four firing rooms, one for each high bay in the VAB, contain control, monitoring and display equipment for automatic vehicle checkout and launch. The display rooms, offices, Launch Information Exchange Facility (LIEF) rooms, and mechanical equipment are located on the fourth floor.

The power demands in this area are large and are supplied by two separate systems, industrial and instrumentation. This division between power systems is designed to protect the instrumentation power

LAUNCH COMPLEX 39

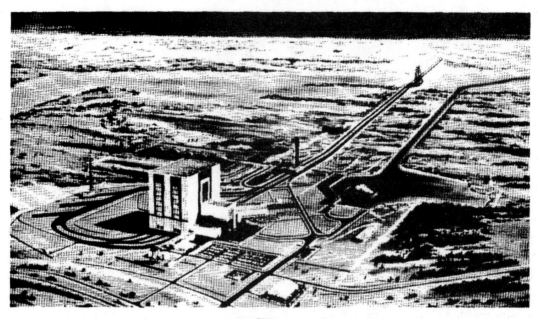

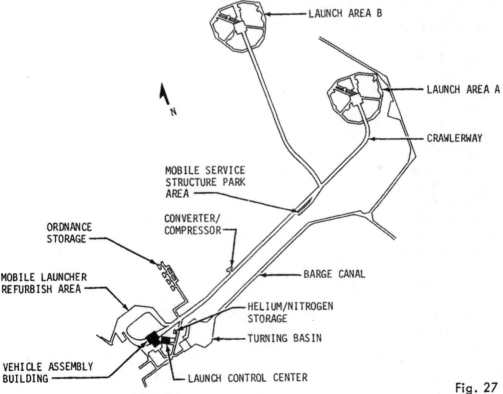

LAUNCH AREA B

LAUNCH AREA A

CRAWLERWAY

MOBILE SERVICE
STRUCTURE PARK
AREA

CONVERTER/
COMPRESSOR

ORDNANCE
STORAGE

BARGE CANAL

MOBILE LAUNCHER
REFURBISH AREA

HELIUM/NITROGEN
STORAGE

TURNING BASIN

VEHICLE ASSEMBLY
BUILDING

LAUNCH CONTROL CENTER

Fig. 27

system from the adverse effects of switching transients, large cycling loads and intermittent motor starting loads. Communication and signal cable troughs extend from the LCC via the enclosed bridge to each ML location in the VAB high bay area. Cableways also connect to the ML refurbishing area and to the Pad Terminal Connection Room (PTCR) at the launch pad. Antennas on the roof provide an RF link to the launch pads and other facilities at KSC.

VEHICLE ASSEMBLY BUILDING

VEHICLE
ASSEMBLY
BUILDING

LAUNCH
CONTROL
CENTER

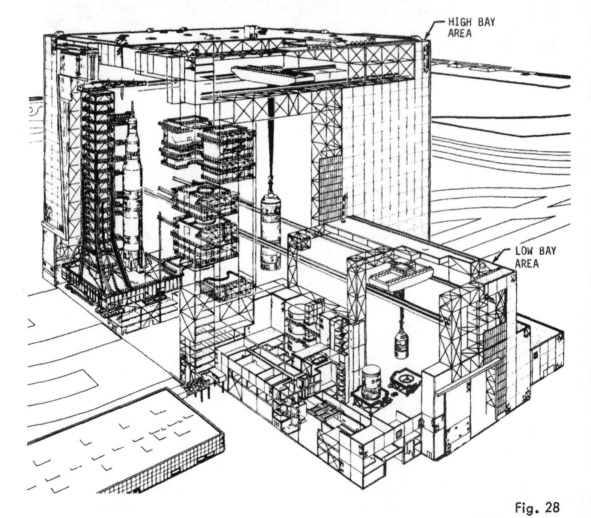

HIGH BAY
AREA

LOW BAY
AREA

Fig. 28

Mobile Launcher

The ML (Figure 29) is a transportable steel structure which, with the C/T, provides the capability to move the erected vehicle to the launch pad. The ML is divided into two functional areas, the launcher base and the umbilical tower. The launcher base is the platform on which a Saturn V vehicle is assembled in the vertical position, transported to a launch site, and launched. The umbilical tower provides access to all important levels of the vehicle during assembly, checkout, and servicing. The equipment used in the servicing, checkout, and launch is installed throughout both the base and tower sections of the ML. The launcher base is a steel structure 25 feet high, 160 feet long, and 135 feet wide. The upper deck, designated level 0, contains, in addition to the umbilical tower, the four hold-down arms and the three tail service masts. There is a 45-foot square opening through the ML base for first stage exhaust.

The base has provisions for attachment to the C/T, six launcher-to-ground mount mechanisms, and four extensible support columns. All electrical/mechanical interfaces between vehicle systems and the VAB or the launch site are located through or adjacent to the base structure. The base houses such items as the computer systems test sets, digital propellant loading equipment, hydraulic test sets, propellant and pneumatic lines, air conditioning and ventilating systems, electrical power systems, and water systems. Fueling operations at the launch area require that the compartments within the structure be pressurized with a supply of uncontaminated air.

The primary electrical power supplied to the ML is divided into four separate services: instrumentation, industrial, in-transit, and emergency. Emergency power is supplied by a diesel-driven generator located in the ground facilities. It is used for obstruction lights, emergency lighting, and for one tower elevator. Water is supplied to the ML for fire, industrial, and domestic purposes.

The umbilical tower is a 380-foot high open steel structure which provides the support for eight umbilical service arms, Apollo Spacecraft access arm, 18 work and access platforms, distribution equipment for the propellant, pneumatic, electrical, and instrumentation subsystems, and other ground support equipment. Two high-speed elevators service 18 landings from level A of the base to the 340-foot tower level. The structure is topped by a 25-ton hammerhead crane. Remote control of the crane is possible from numerous locations on the ML.

The four holddown arms (Figure 30) are mounted on the ML deck, 90° apart around the vehicle base. They position and hold the vehicle on the ML during the VAB checkout, movement to the pad, and pad checkout. The vehicle base is held with a pre-loaded force of 700,000 pounds at each arm. At engine ignition, the vehicle is restrained until proper engine thrust is achieved. The unlatching interval for the four arms should not exceed 0.050 second. If any of the separators fail to operate in 0.180 second, release is effected by detonating an explosive nut link. At launch, the holddown arms quickly release, but the vehicle is prevented from accelerating too rapidly by the controlled-release mechanisms (Figure 30). Each controlled-release mechanism basically consists of a tapered pin inserted in a die which is coupled to the vehicle. Upon vehicle release, the tapered pin is drawn through the die during the first six inches of vehicle travel. There are provisions for as many as 16 mechanisms per vehicle. The precise number is determined on a mission basis.

The three Tail Service Mast (TSM) assemblies (Figure 30) support service lines to the S-IC stage and provide a means for rapid retraction at vehicle lift-off. The TSM assemblies are located on level 0 of the ML base. Each TSM is a counterbalanced structure which is pneumatically/electrically controlled and hydraulically operated. Retraction of the umbilical carrier and vertical rotation of the mast is accomplished simultaneously to ensure no physical contact between the vehicle and mast. The carrier is protected by a hood which is closed by a separate hydraulic system after the mast rotates.

The nine service arms provide access to the space vehicle and support the service lines that are required to sustain the vehicle, as described in Figure 31. The service arms are designated as either pre-flight or in-flight arms. The preflight arms are retracted and locked against the umbilical tower prior to lift-off. The in-flight arms retract at vehicle lift-off. Carrier withdrawal and arm retraction is accomplished by pneumatic and/or hydraulic systems.

MOBILE LAUNCHER

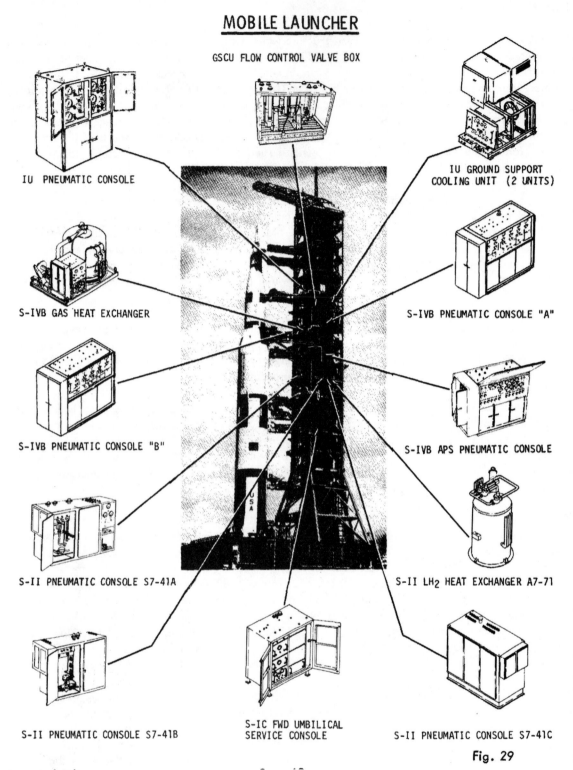

GSCU FLOW CONTROL VALVE BOX

IU PNEUMATIC CONSOLE

IU GROUND SUPPORT
COOLING UNIT (2 UNITS)

S-IVB GAS HEAT EXCHANGER

S-IVB PNEUMATIC CONSOLE "A"

S-IVB PNEUMATIC CONSOLE "B"

S-IVB APS PNEUMATIC CONSOLE

S-II PNEUMATIC CONSOLE S7-41A

S-II LH$_2$ HEAT EXCHANGER A7-71

S-II PNEUMATIC CONSOLE S7-41B

S-IC FWD UMBILICAL
SERVICE CONSOLE

S-II PNEUMATIC CONSOLE S7-41C

Fig. 29

Launch Pad

The launch pad (Figure 32) provides a stable foundation for the ML during Apollo Saturn V launch and pre-launch operations and an interface to the ML for ML and vehicle systems. There are presently two pads at LC 39 located approximately three miles from the VAB area - Each launch site is approximately 3000 feet across.

HOLDDOWN ARMS/TAIL SERVICE MAST

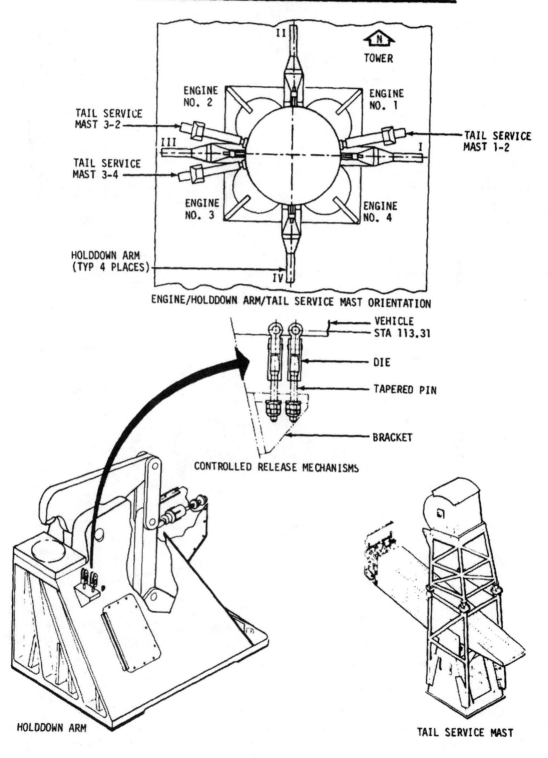

ENGINE/HOLDDOWN ARM/TAIL SERVICE MAST ORIENTATION

CONTROLLED RELEASE MECHANISMS

HOLDDOWN ARM

TAIL SERVICE MAST

Fig. 30

MOBILE LAUNCHER SERVICE ARMS

1 S-IC Intertank (preflight). Provides lox fill and drain interfaces. Umbilical withdrawal by pneumatically driven compound parallel linkage device. Arm may be reconnected to vehicle from LCC. Retract time is 8 seconds. Reconnect time is approximately 5 minutes.

2 S-IC Forward (preflight). Provides pneumatic, electrical, and air-conditioning interfaces. Umbilical withdrawal by pneumatic disconnect in conjunction with pneumatically driven block and tackle/lanyard device. Secondary mechanical system. Retracted at T-20 seconds. Retract time is 8 seconds.

3 S-II Aft (preflight). Provides access to vehicle. Arm retracted prior to liftoff as required.

4 S-II Intermediate (inflight). Provides LH$_2$ and lox transfer, vent line, pneumatic, instrument cooling, electrical, and air-conditioning interfaces. Umbilical withdrawal systems same as S-IVB Forward with addition of a pneumatic cylinder actuated lanyard system. This system operates if primary withdrawal system fails. Retract time is 6.4 seconds (max).

5 S-II Forward (inflight). Provides GH$_2$ vent, electrical, and pneumatic interfaces. Umbilical withdrawal systems same as S-IVB Forward. Retract time is 7.4 seconds (max).

6 S-IVB Aft (inflight). Provides LH$_2$ and lox transfer, electrical, pneumatic, and air-conditioning interfaces. Umbilical withdrawal systems same as S-IVB Forward. Also equipped with line handling device. Retract time is 7.7 seconds (max).

7 S-IVB Forward (inflight). Provides fuel tank vent, electrical, pneumatic, air-conditioning, and preflight conditioning interfaces. Umbilical withdrawal by pneumatic disconnect in conjunction with pneumatic/hydraulic redundant dual cylinder system. Secondary mechanical system. Arm also equipped with line handling device to protect lines during withdrawal. Retract time is 8.4 seconds (max).

8 Service Module (inflight). Provides air-conditioning, vent line, coolant, electrical, and pneumatic interfaces. Umbilical withdrawal by pneumatic/mechanical lanyard system with secondary mechanical system. Retract time is 9.0 seconds (max).

9 Command Module Access Arm (preflight). Provides access to spacecraft through environmental chamber. Arm may be retracted or extended from LCC. Retracted 12° park position until T-4 minutes. Extend time is 12 seconds from this position.

Fig. 31

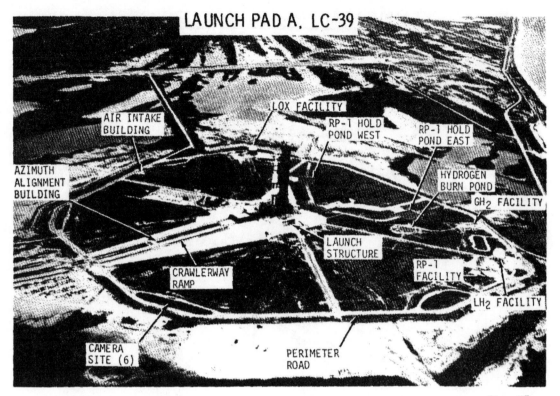

Fig. 32

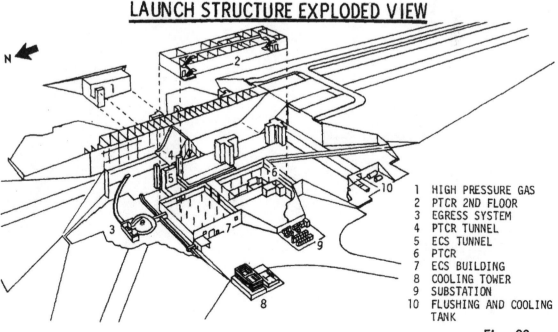

LAUNCH STRUCTURE EXPLODED VIEW

1 HIGH PRESSURE GAS
2 PTCR 2ND FLOOR
3 EGRESS SYSTEM
4 PTCR TUNNEL
5 ECS TUNNEL
6 PTCR
7 ECS BUILDING
8 COOLING TOWER
9 SUBSTATION
10 FLUSHING AND COOLING
   TANK

Fig. 33

LAUNCH PAD INTERFACE SYSTEM

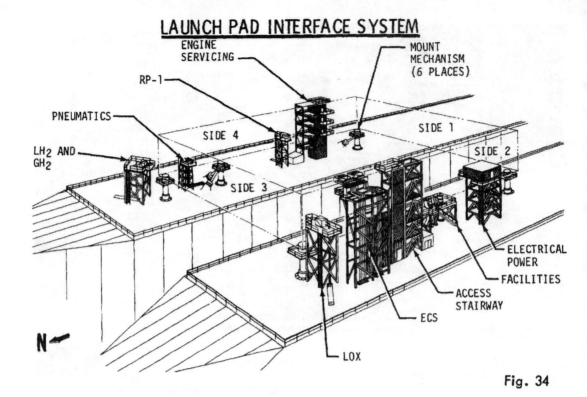

Fig. 34

ELEVATOR /TUBE EGRESS SYSTEM

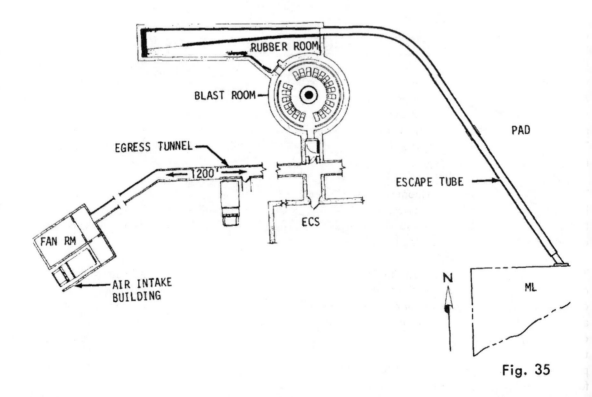

Fig. 35

SLIDE WIRE/CAB EGRESS SYSTEM

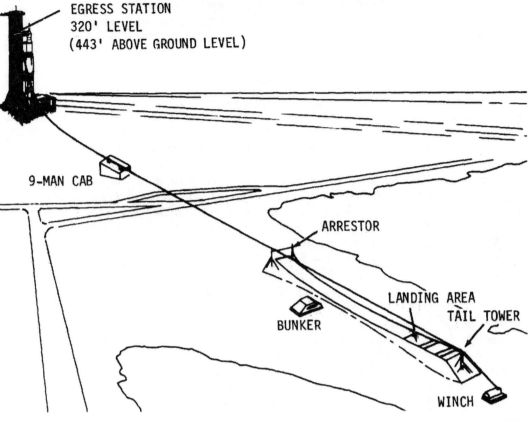

EGRESS STATION
320' LEVEL
(443' ABOVE GROUND LEVEL)

9-MAN CAB

ARRESTOR

LANDING AREA
TAIL TOWER

BUNKER

WINCH

Fig. 36

The launch pad is a cellular, reinforced concrete structure with a top elevation of 42 feet above grade elevation. Located within the fill under the west side of the structure (Figure 33) is a two-story concrete building to house environmental control and pad terminal connection equipment. On the east side of the structure within the fill, is a one-story concrete building to house the high-pressure gas storage battery. On the pad surface are elevators, staircase, and interface structures to provide service to the ML and the MSS. A ramp with a five percent grade provides access from the crawlerway. This is used by the C/T to position the ML/Saturn V and the MSS on the support pedestals. The azimuth alignment building is located on the approach ramp in the crawlerway median strip. A flame trench 58 feet wide by 450 feet long bisects the pod. This trench opens to grade at the north end. The 700,000 pound, mobile, wedge-type flame deflector is mounted on rails in the trench.

The Pad Terminal Connection Room (PTCR) (Figure 33) provides the terminals for communication and data link transmission connections between the ML or MSS and the launch area facilities and between the ML or MSS and the LCC. This facility also accommodates the electronic equipment that simulates functions for checkout of the facilities during the absence of the launcher and vehicle.

The Environmental Control System (ECS) room, located in the pad fill west of the pad structure and north of the PTCR (Figure 33), houses the equipment which furnishes temperature and/or humidity-controlled air or nitrogen for space vehicle cooling at the pad. The ECS room is 96 feet wide by 112 feet long and houses air and nitrogen handling units, liquid chillers, air compressors, a 3000-gallon water-glycol storage tank, and other auxiliary electrical and mechanical equipment. The high-pressure gas storage facility at the pad provides the launch vehicle with high-pressure helium and nitrogen. The launch pad interface system (Figure 34) provides mounting support pedestals for the ML and MSS, an engine access platform, and support structures for fueling, pneumatic, electric power, and environmental control interfaces.

Apollo Emergency Ingress/Egress and Escape System

The Apollo emergency ingress/egress and escape system provides access to and from the Command Module (CM) plus an escape route and safe quarters for the astronauts and service personnel in the event of a serious malfunction prior to launch. The system includes the CM Access Arm, two 600-feet per minute elevators from the 340 foot level to level A of the ML, pad elevator No. 2, personnel carriers located adjacent to the exit of pad elevator No. 2, the escape tube, and the blast room.

The CM Access Arm provides a passage for the astronauts and service personnel from the spacecraft to the 320-foot level of the tower. Egressing personnel take the high speed elevators to level A of the ML, proceed through the elevator vestibule and corridor to pad elevator No. 2, move down this elevator to the bottom of the pad, and enter armored personnel carriers which remove them from the pad area.

When the state of the emergency allows no time for retreat by motor vehicle, egressing personnel, upon reaching level A of the ML, slide down the escape tube into the blast room vestibule, commonly called the "rubber room" (Figure 35). Entrance to the blast room is gained through blast-proof doors controllable from either side. The blast room floor is mounted on coil springs to reduce outside acceleration forces to between 3 and 5 g's. Twenty people may be accommodated for 24 hours. Communication facilities are provided in the room, including an emergency RF link. An underground air duct from the vicinity of the blast room to the remote air intake facility permits egress from the pad structure to the pad perimeter. Provision is made to decrease air velocity in the duct to allow personnel movement through the duct.

An alternate emergency egress system (Figure 36) is referred to as the "Slide Wire." The system consists of a winch-tensioned cable extending from above the 320-foot level of the ML to a 30-foot tail tower on the ground approximately 2200 feet (horizontal projection) from the launcher. A nine-man, tubular-frame cab is suspended from the cable by two brake-equipped trolleys. The unmanned weight of the cab is 1200 pounds and it traverses the distance to the "landing area" in 40 seconds. The cab is decelerated by the increasing drag of a chain attached to a picked-up arresting cable. The occupants of the cab then take refuge in a bunker constructed adjacent to the landing area. The cable has a minimum breaking strength of 53.2 tons and is varied in tension between 18,000 and 32,000 pounds by the winch located beyond the tail tower. The lateral force exerted by the tensioned cable on the ML is negligible relative to the mass of the launcher and the rigidity of the ML tower precludes any effect on tolerances or reliability of tower mechanisms.

Fuel System Facilities

The RP-1 facility consists of three 86,000-gallon steel storage tanks, a pump house, a circulating pump, a transfer pump, two filter-separators, an 8-inch stainless steel transfer line, RP-1 foam generating building, and necessary valves, piping, and controls. Two RP-1 holding ponds (Figure 32), 150 feet by 250 feet, with a water depth of two feet, are located north of the launch pad, one on each side of the north-south axis. The ponds retain spilled RP-1 and discharge water to drainage ditches.

The LH_2 facility (Figure 32) consists of one 850,000-gallon spherical storage tank, a vaporizer/heat exchanger which is used to pressurize the storage tank to 65 psi, a vacuum jacketed, 10-inch invar transfer line and a burn pond venting system. Internal tank pressure provides the proper flow of LH_2 from the storage tank to the vehicle without using a transfer pump. Liquid hydrogen boil-off from the storage and ML areas is directed through vent-piping to bubble-capped headers submerged in the burn pond where a hot wire ignition system maintains the burning process.

LOX System Facility

The LOX facility (Figure 32) consists of one 900,000-gallon spherical storage tank, a LOX vaporizer to pressurize the storage tank, main fill and replenish pumps, a drain basin for venting and dumping of LOX, and two transfer lines.

Azimuth Alignment Building

The azimuth alignment building (Figure 32) houses the auto-collimator theodolite which senses, by a light source, the rotational output of the stable platform in the Instrument Unit of the launch vehicle. This instrument monitors the critical inertial reference system prior to launch.

Photography Facilities

These facilities support photographic camera and closed circuit television equipment to provide real-time viewing and photographic documentation coverage. There are six camera sites in the launch pad area. These sites cover pre-launch activities and launch operations from six different angles at a radial distance of approximately 1300 feet from the launch vehicle. Each site has four engineering, sequential cameras and one fixed, high-speed metric camera.

Pad Water System Facilities

The pad water system facilities furnish water to the launch pad area for fire protection, cooling, and quenching. Specifically, the system furnishes water for the industrial water system, flame deflector cooling and quench, ML deck cooling and quench, ML tower fogging and service arm quench, sewage treatment plant, Firex water system, liquid propellant facilities, ML and MSS fire protection, and all fire hydrants in the pad area.

Mobile Service Structure

The MSS (Figure 37) provides access to those portions of the space vehicle which cannot be serviced from the ML while at the launch pad. The MSS is transported to the launch site by the C/T where it is used during launch pad operations. It is removed from the pad a few hours prior to launch and returned to its parking area 7000 feet from the nearest launch pad. The MSS is approximately 402 feet high and weighs 12 million pounds. The tower structure rests on a base 135 feet by 135 feet. At the top, the tower is 87 feet by 113 feet.

The structure contains five work platforms which provide access to the space vehicle. The outboard sections of the platforms open to accept the vehicle and close around it to provide access to the launch vehicle and spacecraft. The lower two platforms are vertically adjustable to serve different parts of the launch vehicle. The upper three platforms are fixed but can be disconnected from the tower and relocated as a unit to serve different vehicle configurations. The second and third platforms from the top are enclosed and provide environmental control for the spacecraft.

The MSS is equipped with the following systems: air conditioning, electrical power, various communication networks, fire protection, compressed air, nitrogen pressurization, hydraulic pressure, potable water, and spacecraft fueling.

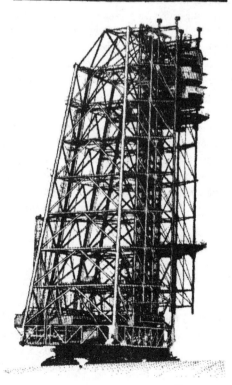

MOBILE SERVICE STRUCTURE

Fig. 37

Crawler-Transporter

The C/T (Figure 38) is used to transport the ML, including the space vehicle, and the MSS to and from the launch pad. The C/T is capable of lifting, transporting, and lowering the ML or the MSS, as required, without the aid of auxiliary equipment. The C/T supplies limited electric power to the ML and the MSS during transit.

The C/T consists of a rectangular chassis which is supported through a suspension system by four dual-tread, crawler-trucks. The overall length is 131 feet and the overall width is 114 feet. The unit weighs approximately six million pounds. The C/T is powered by self-contained, diesel electric generator units. Electric motor driven pumps provide hydraulic power for steering and suspension control. Air conditioning and ventilation are provided where required.

CRAWLER TRANSPORTER

The C/T can be operated with equal facility in either direction. Control cabs are located at each end. The leading cab, in the direction of travel, has complete control of the vehicle. The rear cab, however, has override controls for the rear tracks only. Maximum C/T speed is 2 mph unloaded, 1 mph with full load on level grade, and 0.5 mph with full load on a five percent grade. It has a 500-foot minimum turning radius and can position the ML or the MSS on the facility support pedestals within ±2inches.

Fig. 38

VEHICLE ASSEMBLY AND CHECKOUT

The Saturn V Launch Vehicle propulsive stages and the IU are, upon arrival at KSC, transported to the VAB by special carriers. The S-IC stage is erected on an ML in one of the checkout bays in the high bay area. The S-II and S-IVB stages and the IU are delivered to preparation and checkout cells in the low bay area for inspection, checkout, and pre-erection preparations. All components of the space vehicle, including the Apollo Spacecraft and Launch Escape System, are then assembled vertically on the ML in the high bay area. Following assembly, the space vehicle is connected to the LCC via a high-speed data link for integrated checkout and a simulated flight test. When checkout is completed, the C/T picks up the ML with the assembled space vehicle and moves it to the launch site via the crawlerway.

At the launch site, the ML is emplaced and connected to system interfaces for final vehicle checkout and launch monitoring. The MSS is transported from its parking area by the C/T and positioned on the side of the vehicle opposite the ML. A flame deflector is moved on its track to its position beneath the blast opening of the ML to deflect the blast from the S-IC stage engines. During the pre-launch checkout, the final system checks are completed, the MSS is removed to the parking area, propellants are loaded, various items of support equipment are removed from the ML, and the vehicle is readied for launch. After vehicle launch, the C/T transports the ML to the parking area near the VAB for refurbishment.

MISSION MONITORING, SUPPORT, AND CONTROL

GENERAL

Mission execution involves the following functions: pre-launch checkout and launch operations; tracking the space vehicle to determine its present and future positions; securing information on the status of the flight crew and space vehicle systems (via telemetry); evaluation of telemetry information; commanding the space vehicle by transmitting real-time and updata commands to the onboard computer; voice communication between flight and ground crews; and recovery operations.

These functions require the use of a facility to assemble and launch the space vehicle (see Launch Complex);

a central flight control facility; a network of remote stations located strategically around the world; a method of rapidly transmitting and receiving information between the space vehicle and the central flight control facility; a realtime data display system in which the data is made available and presented in usable form at essentially the same time that the data event occurred; and ships/aircraft to recover the spacecraft on return to earth.

The flight crew and the following organizations and facilities participate in mission control operations:

1. Mission Control Center (MCC), Manned Spacecraft Center (MSC), Houston, Texas. The MCC contains the communication, computer, display, and command systems to enable the flight controllers to effectively monitor and control the space vehicle.

2. Kennedy Space Center (KSC), Cape Kennedy, Florida. The space vehicle is launched from KSC and controlled from the Launch Control Center (LCC), as described previously. Pre-launch, launch, and powered flight data are collected at the Central Instrumentation Facility (CIF) at KSC from the launch pads, CIF receivers, Merritt Island Launch Area (MILA), and the down-range Air Force Eastern Test Range (AFETR) stations. This data is transmitted to MCC via the Apollo Launch Data System (ALDS). Also located at KSC (ETR) is the Impact Predictor (IP), for range safety purposes.

3. Goddard Space Flight Center (GSFC), Greenbelt, Maryland. GSFC manages and operates the Manned Space Flight Network (MSFN) and the NASA communications (NASCOM) networks. During flight, the MSFN is under operational control of the MCC.

4. George C. Marshall Space Flight Center (MSFC), Huntsville, Alabama. MSFC, by means of the Launch Information Exchange Facility (LIEF) and the Huntsville Operations Support Center (HOSC) provides launch vehicle systems real-time support to KSC and MCC for pre-flight, launch, and flight operations. A block diagram of the basic flight control interfaces is shown in Figure 39.

BASIC TELEMETRY, COMMAND, AND COMMUNICATION INTERFACES FOR FLIGHT CONTROL

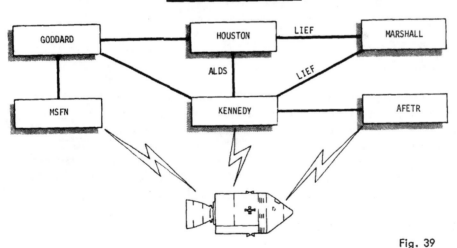

Fig. 39

VEHICLE FLIGHT CONTROL CAPABILITY

Flight operations are controlled from the MCC. The MCC has two flight control rooms. Each control room, called a Mission Operations Control Room (MOCR), is used independently of the other and is capable of controlling individual Staff Support Rooms (SSR's) located adjacent to the MOCR. The SSR's are manned by flight control specialists who provide detailed support to the MOCR. Figure 40 outlines the organization of the MCC for flight control and briefly describes key responsibilities. Information flow within the MOCR is shown in Figure 41.

MCC ORGANIZATION

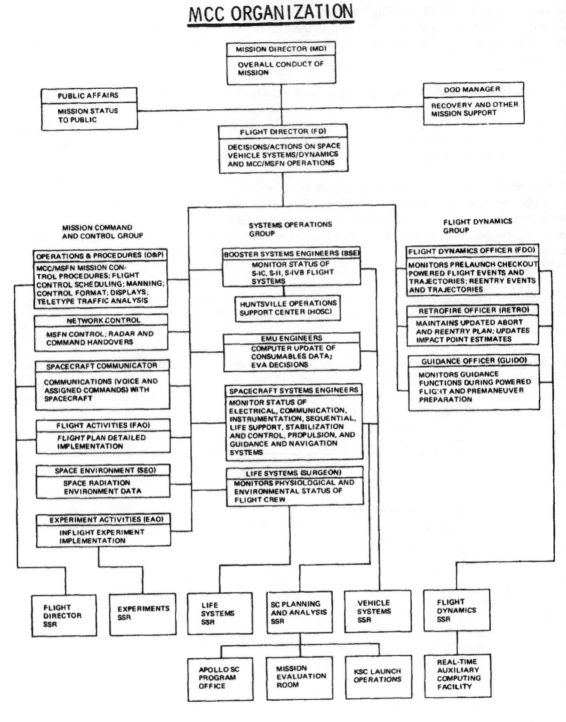

Fig. 40

The consoles within the MOCR and SSR's permit the necessary interface between the flight controllers and the spacecraft. The displays and controls on these consoles and other group displays provide the capability to monitor and evaluate data concerning the mission and, based on these evaluations, to recommend or take appropriate action on matters concerning the flight crew and spacecraft.

INFORMATION FLOW MISSION OPERATIONS CONTROL ROOM

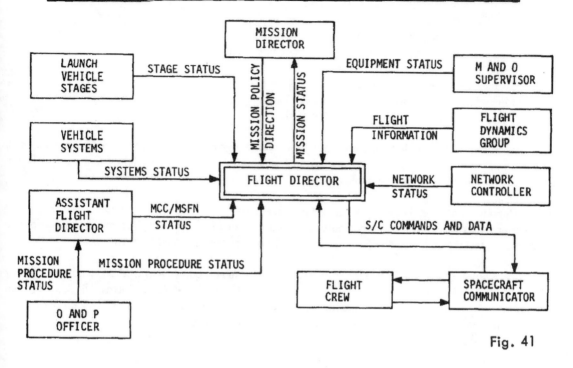

Fig. 41

MCC FUNCTIONAL CONFIGURATION

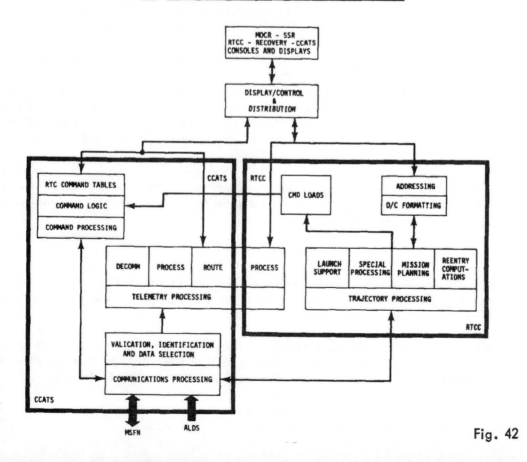

Fig. 42

Problems concerning crew safety and mission success are identified to flight control personnel in the following ways:

1. Flight crew observations;
2. Flight controller real-time observations;
3. Review of telemetry data received from tape recorder playback;
4. Trend analysis of actual and predicted values;
5. Review of collected data by systems specialists;
6. Correlation and comparison with previous mission data;
7. Analysis of recorded data from launch complex testing.

The facilities at the MCC include an input/output processor designated as the Command, Communications, and Telemetry System (CCATS) and a computational facility, the RealTime Computer Complex (RTCC). Figure 42 shows the MCC functional configuration.

The CCATS consists of three Univac 494 general purpose computers. Two of the computers are configured so that either may handle all of the input/output communications for two complete missions. One of the computers acts as a dynamic standby. The third computer is used for non mission activities.

The RTCC is a group of five IBM 360 large-scale, general purpose computers. Any of the five computers may be designated as the Mission Operations Computer (MOC). The MOC performs all the required computations and display formatting for a mission. One of the remaining computers will be a dynamic standby. Another pair of computers may be used for a second mission or simulation.

Space Vehicle Tracking

From lift-off of the launch vehicle to insertion into orbit, accurate position data are required to allow the Impact Predictor (IP) to function effectively as a Range Safety device, and the RTCC to compute a trajectory and an orbit. These computations are required by the flight controllers to evaluate the trajectory, the orbit, and/or any abnormal situations to ensure safe recovery of the astronauts. The launch tracking data are transmitted from the AFETR site to the IP and thence to the RTCC via high speed data communications circuits. The IP also generates spacecraft inertial positions and inertial rates of motion in real-time.

During boost the trajectory is calculated and displayed on consoles and plot boards in the MOCR and SSR's. Also displayed are telemetry data concerning status of launch vehicle and spacecraft systems. If the space vehicle deviates excessively from the nominal flight path, or if any critical vehicle condition exceeds tolerance limits, or if the safety of the astronauts or range personnel is endangered, a decision is made to abort the mission.

During the orbit phase of a mission, all stations that are actively tracking the spacecraft will transmit the tracking data through GSFC to the RTCC by teletype. If a thrusting maneuver is performed by the spacecraft, high-speed tracking data is also transmitted.

Command System

The Apollo ground command systems have been designed to work closely with the telemetry and trajectory systems to provide flight controllers with a method of "closed loop" command. The astronauts and flight controllers act as links in this operation.

To prevent spurious commands from reaching the space vehicle, switches on the Command Module console block uplink data from the onboard computers. At the appropriate times, the flight crew will move the switches from the "BLOCK" to the "ACCEPT" positions and thus permit the flow of uplink data.

With a few exceptions, commands to the space vehicle fall into two categories: realtime commands, and command loads (also called computer loads, computer update, loads, or update).

Real-time commands are used to control space vehicle systems or subsystems from the ground. The execution of a real-time command results in immediate reaction by the affected system. Real-time commands are stored prior to the mission in the Command Data Processor (CDP) at the applicable command site. The CDP, a Univac 642B, general-purpose digital computer, is programmed to format, encode, and output commands when a request for uplink is generated.

Command loads are generated by the real-time computer complex on request of flight controllers. Command loads are based on the latest available telemetry and/or trajectory data. Flight controllers typically required to generate a command load include the Booster Systems Engineer (BSE), the Flight Dynamics Officer (FDO), the Guidance Officer (GUIDO), and the Retrofire Officer (RETRO).

Display and Control System

The MCC is equipped with facilities which provide for the input of data from the MSFN and KSC over a combination of high-speed data, low-speed data, wide-band data, teletype, and television channels. These data are computer processed for display to the flight controllers.

Several methods of displaying data are used including television (projection TV, group displays, closed circuit TV, and TV monitors), console digital readouts, and event lights. The display and control system interfaces with the RTCC and includes computer request, encoder multiplexer, plotting display, slide file, digital-to-TV converter, and telemetry event driver equipment.

A control system is provided for flight controllers to exercise their respective functions for mission control and technical management. This system is comprised of different groups of consoles with television monitors, request keyboards, communications equipment, and assorted modules added as required to provide each operational position in the MOCR with the control and display capabilities required for the particular mission.

CONTINGENCY PLANNING AND EXECUTION

Planning for a mission begins with the receipt of mission requirements and objectives. The planning activity results in specific plans for pre-launch and launch operations, pre-flight training and simulation, flight control procedures, flight crew activities, MSFN and MCC support, recovery operations, data acquisition and flow, and other mission-related operations. Numerous simulations are planned and performed to test procedures and train flight control and flight crew teams in normal and contingency operations.

MCC Role in Aborts

After launch and from the time the space vehicle clears the ML, the detection of slowly-deteriorating conditions which could result in an abort is the prime responsibility of MCC; prior to this time, it is the prime responsibility of LCC. In the event such conditions are discovered, MCC requests abort of the mission or, circumstances permitting, sends corrective commands to the vehicle or requests corrective flight crew actions. In the event of a non catastrophic contingency, MCC recommends alternate flight procedures, and mission events are rescheduled to derive maximum benefit from the modified mission.

VEHICLE FLIGHT CONTROL PARAMETERS

In order to perform flight control monitoring functions, essential data must be collected, transmitted, processed, displayed, and evaluated to determine the space vehicle's capability to start or continue the mission.

Parameters Monitored by LCC

The launch vehicle checkout and pre-launch operations monitored by the Launch Control Center (LCC) determine the state of readiness of the launch vehicle, ground support, telemetry, range safety, and other operational support systems. During the final countdown, hundreds of parameters are monitored to ascertain vehicle, system, and component performance capabilities. Among these parameters are the "redlines." The redline values must be within the predetermined limits or the countdown will be halted. In addition to the redlines, there are a number of operational support elements such as ALDS, range instrumentation, ground tracking and telemetry stations, and ground support facilities which must be operational at specified times in the countdown.

Parameters Monitored by Booster Systems Group

The Booster Systems Group (BSG) monitors launch vehicle systems (S-IC, S-II, S-IVB, and IU) and advises the flight director and flight crew of any system anomalies. It is responsible for confirming in-flight power, stage ignition, holddown release, all engines go, engine cutoffs, etc. BSG also monitors attitude control, stage separations, and digital commanding of LV systems.

Parameters Monitored by Flight Dynamics Group

The Flight Dynamics Group monitors and evaluates the powered flight trajectory and makes the abort decisions based on trajectory violations. It is responsible for abort planning, entry time and orbital maneuver determinations, rendezvous planning, inertial alignment correlation, landing point prediction, and digital commanding of the guidance systems.

The MOCR positions of the Flight Dynamics Group include the Flight Dynamics Officer (FDO), the Guidance Officer (GUIDO), and the Retrofire Officer (RETRO). The MOCR positions are given detailed, specialized support by the Flight Dynamics SSR.

The surveillance parameters measured by the ground tracking stations and transmitted to the MCC are computer processed into plot board and digital displays. The Flight Dynamics Group compares the actual data with pre-mission, calculated, nominal data and is able to determine mission status.

Parameters Monitored by Spacecraft Systems Group

The Spacecraft Systems Group monitors and evaluates the performance of spacecraft electrical, optical, mechanical, and life support systems; maintains and analyzes consumables status; prepares the mission log; coordinates telemetry playback; determines spacecraft weight and center of gravity; and executes digital commanding of spacecraft systems.

The MOCR positions of this group include the Command and Service Module Electrical, Environmental, and Communications Engineer (CSM EECOM), the CSM Guidance, Navigation, and Control Engineer (CSM GNC), the Lunar Module Electrical, Environmental, and Communications Engineer (LM EECOM), and the LM Guidance, Navigation, and Control Engineer (LM GNC). These positions are backed up with detailed support from the Vehicle Systems SSR.

Parameters Monitored by Life Systems Group

The Life Systems Group is responsible for the well-being of the flight crew. The group is headed by the Flight Surgeon in the MOCR. Aeromedical and environmental control specialists in the Life Systems SSR provide detailed support to the Flight Surgeon. The group monitors the flight crew health status and environmental/biomedical parameters.

MANNED SPACE FLIGHT NETWORK

The Manned Space Flight Network (MSFN) (Figure 43) is a global network of ground stations, ships, and aircraft designed to support manned and unmanned space flights. The network provides tracking, telemetry, voice and teletype communications, command, recording, and television capabilities. The network is specifically configured to meet the requirements of each mission.
MSFN stations are categorized as lunar support stations (deep-space tracking in excess of 15,000 miles), near-space support stations with Unified S-Band (USB) equipment, and near-space support stations without

USB equipment. The deep-space S-band capability is attained with 85-foot antennas located at: Honeysuckle Creek, Australia; Goldstone, California; and Madrid, Spain. MSFN stations include facilities operated by NASA, the United States Department of Defense (DOD), and the Australian Department of Supply (DOS). The DOD facilities include the Eastern Test Range (ETR), Western Test Range (WTR), White Sands Missile Range (WSMR), Range Instrumentation Ships (RIS), and Apollo Range Instrumentation Aircraft (ARIA).

The MSFN coverage by ground stations is supplemented by the five Range Instrumentation Ships. The number and position of the ships is determined for each mission. The Vanguard, Redstone, and Mercury support earth-orbital insertion and translunar injection phases of a mission. The Huntsville and Watertown support reentry phases of a mission. The ships operate as integral stations of the MSFN, meeting target acquisition, tracking, telemetry, communications, and command and control requirements. The reentry ships have no telemetry computer, command control system, mission control center, or satellite communications terminal. The DOD operates the ships in support of NASA/DOD missions with an Apollo priority. The Military Sea Transport Service provides the maritime crew and the WTR provides the instrumentation crews by contract. The WTR also has the operational management responsibility for the ships. The ships may contribute to the recovery phase as necessary for contingency landings.

Eight modified C-135 aircraft supplement the ground stations and instrumentation ships as highly mobile "gap fillers." The ARIA support other space and missile projects when not engaged in their primary mission of Apollo support. The ARIA provide two-way relay of; voice communications between the spacecraft and surface stations; reception, recording, and retransmission of telemetry signals from the spacecraft to the ground (postpass). The aircraft are used shortly before, during, and shortly after injection burn; from initial communications blackout to final landing; for coverage of a selected abort area in the event of a "no-go" decision after injection or for any irregular reentry. The ARIA have an endurance of about 10 hours and a cruise airspeed of about 450 knots.

NASA COMMUNICATIONS NETWORK

The NASA Communications (NASCOM) network (Figure 44) is a point-to-point communications system connecting the MSFN stations to the MCC. NASCOM is managed by the Goddard Space Flight Center, where the primary communications switching center is located. Three smaller NASCOM switching centers are located at London, Honolulu, and Canberra. Patrick AFB, Florida and Wheeler AFB, Hawaii serve as switching centers for the DOD eastern and western test ranges, respectively. The MSFN stations throughout the world are interconnected by landline, undersea cable, radio, and communications satellite circuits. These circuits carry teletype, voice, and data in real-time support of the missions.

Each MSFN USB land station has a minimum of five voice/data circuits and two teletype circuits. The Apollo insertion and injection ships have a similar capability through the communications satellites.

ACN	ASCENSION IS. (NASA STATION)	HSK	HONEYSUCKLE CR. AUST.
ACSW	CANBERRA SWITCHING STA.	HTV	USNS HUNTSVILLE
ANG	ANTIGUA ISLAND	LLDN	LONDON SWITCHING CENTER
ANT	AFETR SITE ANTIGUA ISLAND	LROB	MADRID, SPAIN SWITCHING CENTER
AOCC	AIRCRAFT OPERATIONS CONTROL CENTER		
MAD	MADRID, SPAIN	ARIA	APOLLO RANGE INSTRUMENTATION AIRCRAFT
MER	USNS MERCURY	BDA	BERMUDA
MCC	MISSION CONTROL CENTER	CAL	CALIFORNIA (VANDENBERG AFB)
MIL	MERRITT ISLAND, FLA.	CDSC	COMMUNICATION DISTRIBUTION SWITCHING CENTER
MSFC	MARSHALL SPACE FLIGHT CENTER		
PGSW	GUAM SWITCHING CENTER	CRO	CARNARVON, AUSTRALIA
PHON	HONOLULU SWITCHING STA.	CYI	GRAND CANARY ISLAND
RED	USNS REDSTONE.	ETR	EASTERN TEST RANGE
TAN	TANANARIVE, MALAGASY	GBM	GRAND BAHAMA IS.
TEX	CORPUS CHRISTI, TEXAS	GDS	GOLDSTONE, CALIFORNIA
VAN	USNS VANGUARD	GSFC	GODDARD SPACE FLIGHT CENTER
WHS	WHITE SANDS, NEW MEXICO	GWM	GUAM
WOM	WOOMERA, AUSTRALIA	GYM	GUAYMAS, MEXICO
WTR	WESTERN TEST RANGE	HAW	HAWAII

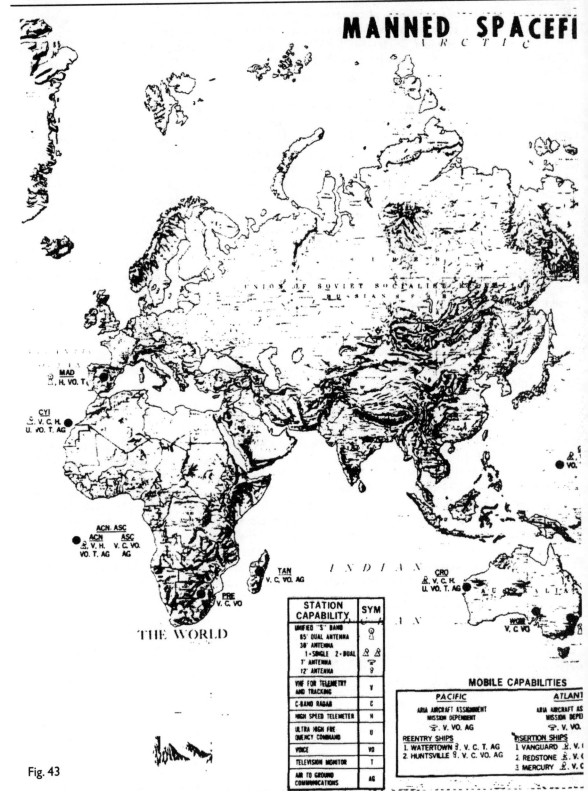

MANNED SPACEFI

Fig. 43

STATION CAPABILITY	SYM
UNIFIED "S" BAND 85' DUAL ANTENNA	
30' ANTENNA 1 - SINGLE 2 - DUAL	
7' ANTENNA	
12' ANTENNA	
VHF FOR TELEMETRY AND TRACKING	V
C-BAND RADAR	C
HIGH SPEED TELEMETER	H
ULTRA HIGH FREQUENCY COMMAND	U
VOICE	VO
TELEVISION MONITOR	T
AIR TO GROUND COMMUNICATIONS	AG

MOBILE CAPABILITIES

PACIFIC	ATLANTI
ARIA AIRCRAFT ASSIGNMENT MISSION DEPENDENT V. VO. AG	ARIA AIRCRAFT ASSIGNMENT MISSION DEPENDENT V. VO.
REENTRY SHIPS 1. WATERTOWN V. C. T. AG 2. HUNTSVILLE V. C. VO. AG	INSERTION SHIPS 1. VANGUARD V. 2. REDSTONE V. 3. MERCURY V. C

APOLLO LAUNCH DATA SYSTEM (ALDS)

The Apollo Launch Data System (ALDS) between KSC and MSC is controlled by MSC and is not routed through GSFC. The ALDS consists of wide-band telemetry, voice coordination circuits, and a high-speed circuit for the Countdown and Status Transmission System (CASTS). In addition, other circuits are provided for launch coordination, tracking data, simulations, public information, television, and recovery.

FLIGHT NETWORK

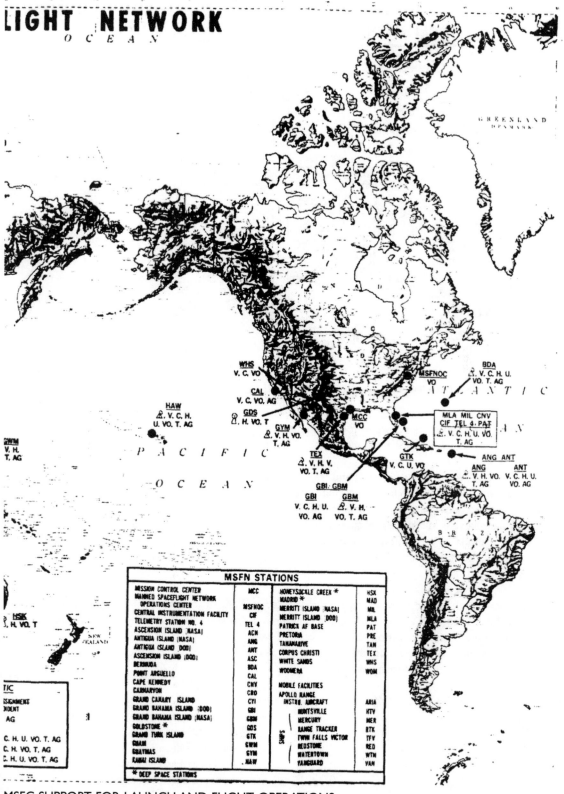

MSFN STATIONS			
MISSION CONTROL CENTER	MCC	HONEYSUCKLE CREEK *	HSK
MANNED SPACEFLIGHT NETWORK OPERATIONS CENTER	MSFNOC	MADRID *	MAD
CENTRAL INSTRUMENTATION FACILITY	CIF	MERRITT ISLAND (NASA)	MIL
TELEMETRY STATION NO. 4	TEL 4	MERRITT ISLAND (DOD)	MLA
ASCENSION ISLAND (NASA)	ACN	PATRICK AF BASE	PAT
ANTIGUA ISLAND (NASA)	ANG	PRETORIA	PRE
ANTIGUA ISLAND (DOD)	ANT	TANANARIVE	TAN
ASCENSION ISLAND (DOD)	ASC	CORPUS CHRISTI	TEX
BERMUDA	BDA	WHITE SANDS	WHS
POINT ARGUELLO	CAL	WOOMERA	WOM
CAPE KENNEDY	CNV		
CARNARVON	CRO	MOBILE FACILITIES	
GRAND CANARY ISLAND	CYI	APOLLO RANGE INSTR. AIRCRAFT	ARIA
GRAND BAHAMA ISLAND (DOD)	GBI	HUNTSVILLE	HTV
GRAND BAHAMA ISLAND (NASA)	GBM	MERCURY	MER
GOLDSTONE *	GDS	RANGE TRACKER	RTK
GRAND TURK ISLAND	GTK	TWIN FALLS VICTOR	TFV
GUAM	GWM	REDSTONE	RED
GUAYMAS	GYM	WATERTOWN	WTN
KAUAI ISLAND	HAW	VANGUARD	VAN

* DEEP SPACE STATIONS

MSFC SUPPORT FOR LAUNCH AND FLIGHT OPERATIONS

The Marshall Space Flight Center (MSFC), by means of the Launch Information Exchange Facility (LIEF) and the Huntsville Operations Support Center (HOSC), provides real-time support of launch vehicle pre-launch, launch, and flight operations. MSFC also provides support, via LIEF, for post-flight data delivery and evaluation.

TYPICAL MISSION COMMUNICATIONS NETWORK

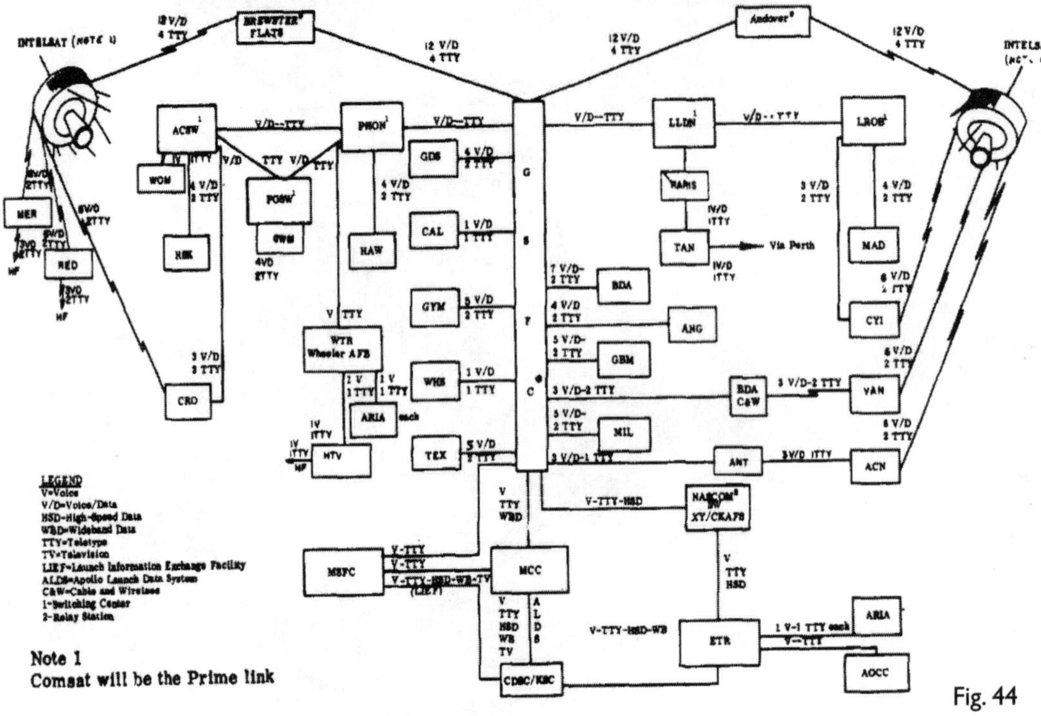

Fig. 44

In-depth real-time support is provided for pre-launch, launch, and flight operations from HOSC consoles manned by engineers who perform detailed system data monitoring and analysis.

Pre-launch flight wind monitoring analysis and trajectory simulations are jointly performed by MSFC and MSC personnel located at MSFC during the terminal countdown. Beginning at T-24 hours, actual wind data is transmitted periodically from KSC to the HOSC. These measurements are used by the MSFC/MSC wind monitoring team in vehicle flight digital simulations to verify the capability of the vehicle with these winds. In the event of marginal wind conditions, contingency data are provided MSFC in real-time via the Central Instrumentation Facility (CIF). DATA-CORE and trajectory simulations are performed on-line to expedite reporting to KSC.

During the pre-launch period, primary support is directed to KSC. At lift-off primary support transfers from KSC to the MCC. The HOSC engineering consoles provide support as required to the Booster Systems Group for S-IVB/IU orbital operations by monitoring detailed instrumentation for the evaluation of system in-flight and dynamic trends, assisting in the detection and isolation of vehicle malfunctions and providing advisory contact with vehicle design specialists.

ABBREVIATIONS AND ACRONYMS

ac	Alternating Current
AEA	Abort Electronics Assembly
AFB	Air Force Base
AFETR	Air Force Eastern Test Range
AGS	Abort Guidance Subsystem
ALDS	Apollo Launch Data System
AM	Amplitude Modulation
AOT	Alignment Optical Telescope
APS	Auxiliary Propulsion System (S-IVB)
APS	Ascent Propulsion System (LM)
ARIA	Apollo Range Instrumentation Aircraft

ARS	Atmosphere Revitalization Subsystem
AS	Apollo Saturn
AS	Ascent Stage (LM)
ASI	Augmented Spark Igniter
BPC	Boost Protective Cover
BSE	Boost Systems Engineer
CASTS	Countdown and Status Transmission System
CCATS	Communications, Command, and Telemetry System
CCS	Command Communications System
CDP	Command Data Processor (MSFN Site)
CES	Control Electronics Subsystem
CIF	Central Instrumentation Facility
CM	Command Module
COAS	Crewman Optical Alignment Sight
CS	Communications System
CSM	Command Service Module
C/T	Crawler/Transporter
CWEA	Caution and Warning Electronics Assembly
CWG	Constant-Wear Garment
DATA-CORE	CIF Telemetry Conversion System
dc	Direct Current
DEDA	Data Entry and Display Assembly
DOD	Department of Defense
DOS	Department of Supply (Australia)
DPS	Descent Propulsion System (LM)
DS	Descent Stage (LM)
DSEA	Data Storage Electronics Assembly
ECS	Environmental Control System
EDS	Emergency Detection System
EDS	Explosive Devices System (LM)
ELS	Earth Landing System
EMS	Entry Monitor System
EMU	Extravehicular Mobility Unit
EPS	Electrical Power System
ETR	Eastern Test Range
EV	Extravehicular
EVA	Extravehicular Activity
FCC	Flight Control Computer (IU, analog)
FDAI	Flight Director Attitude Indicator
FDO	Flight Dynamics Officer
g	Gravity force at sea level (1 g)
GDC	Gyro Display Coupler
GH_2	Gaseous Hydrogen
GN_2	Gaseous Nitrogen
GNCS	Guidance, Navigation, and Control System
GN&CS	Guidance, Navigation, and Control System (LM)
GOX	Gaseous Oxygen
GSE	Ground Support Equipment
GUIDO	Guidance Officer
GSFC	Goddard Space Flight Center
H_2	Hydrogen
HF	High Frequency
HOSC	Huntsville Operations Support Center

HTS	Heat Transport Subsystem (LM)
ICG	In-flight Coverall Garment
IMU	Inertial Measurement Unit
IP	Impact Predictor (of KSC)
IS	Instrumentation System (LM)
IU	Instrument Unit
KSC	Kennedy Space Center
LC	Launch Complex
LCC	Launch Control Center
LCG	Liquid-Cooling Garment
LEA	Launch Escape Assembly
LEB	Lower Equipment Bay
LES	Launch Escape System
LET	Launch Escape Tower
LH	Liquid Hydrogen
LIEF	Launch Information Exchange Facility
LM	Lunar Module
LN_2	Liquid Nitrogen
LOX, LO_2	Liquid Oxygen
LR	Landing Radar
LV	Launch Vehicle
LVDA	Launch Vehicle Data Adapter
LVDC	Launch Vehicle Digital Computer
MCC	Mission Control Center
MILA	Merritt Island Launch Area
ML	Mobile Launcher
MMH	Monomethyl Hydrazine
MOC	Mission Operations Computer
MOCR	Mission Operations Control Room
MSC	Manned Spacecraft Center
MSFC	Marshall Space Flight Center
MSFN	Manned Space Flight Network
MSS	Mobile Service Structure
NASCOM	NASA Communications Network
N_2H_4	Hydrazine
N_2O_4	Nitrogen Tetroxide
NPSH	Net Positive Suction Head
O_2	Oxygen
OMR	Operations Management Room
OPS	Oxygen Purge System
OSCPCS	Oxygen Supply and Cabin Pressure Control Subsystem
OSR	Operations Support Room
PCMTEA	Pulse-Code-Modulation and Timing Electronics Assembly
PDS	Propellant Dispersion System
PGA	Pressure Garment Assembly
PGNCS	Primary Guidance Navigation and Control System (LM)
PGNS	Primary Guidance and Navigation Subsystem (LM)
PLSS	Portable Life Support System
PTCR	Pad Terminal Connection Room
PU	Propellant Utilization

RCS	Reaction Control System
RETRO	Direction Opposite to Velocity Vector
RF	Radio Frequency
RIS	Range Instrumentation Ship
RP-1	Rocket Propellant (refined kerosene)
RR	Rendezvous Radar
RTCC	Real Time Computer Complex
SC	Spacecraft
SCS	Stabilization and Control System
SCEA	Signal Conditioning Electronics Assembly
SECS	Sequential Events Control System
SLA	Spacecraft LM Adapter
SM	Service Module
SPS	Service Propulsion System
SSR	Staff Support Room
SV	Space Vehicle
TCA	Thrust Chamber Assembly
TCS	Thermal Conditioning System
TSM	Tail Service Mast
TV	Television
UDMH	Unsymmetrical Dimethylhydrazine
USB	Unified S-band
UHF	Ultra-High Frequency
VAB	Vehicle Assembly Building
VHF	Very High Frequency
WMS	Water Management Subsystem (LM)
WSMR	White Sands Missile Range
WTR	Western Test Range

POST LAUNCH MISSION OPERATION REPORT

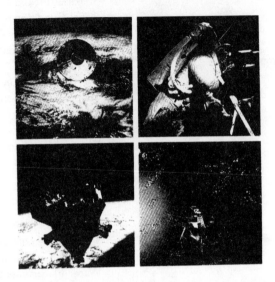

APOLLO 9 (AS-504) MISSION

MEMORANDUM   6 May 1969

To: A/Administrator

From: MA/Apollo Program Director

Subject: Apollo 9 Mission (AS-504) Post Launch Report #1

The Apollo 9 mission was successfully launched from the Kennedy Space Center on Monday, 3 March 1969 and was completed as planned, with recovery of the spacecraft and crew in the Atlantic recovery area on Thursday, 13 March 1969. Initial evaluation of the flight, based upon quick-look data and crew debriefing, indicates that all mission objectives were attained. Further detailed analysis of all data is continuing and appropriate refined results of the mission will be reported in Manned Space Flight Center technical reports.

Based on the mission performance as described in this report, I am recommending that the Apollo 9 mission be adjudged as having achieved agency preset primary objectives and be considered a success.

Sam C. Phillips
Lt. General, USAF
Apollo Program Director

APPROVAL

George E. Mueller
Associate Administrator for Manned Space Flight

GENERAL

The Apollo 9 (AS-504) mission was the first manned flight involving the Lunar Module. The crew were James A. McDivitt, Commander; David R. Scott, Command Module Pilot; and Russell L. Schweickart, Lunar Module Pilot. Launch had been initially scheduled for 28 February 1969, but was postponed for three days because all three crewmen had virus respiratory infections. The countdown was accomplished without any unscheduled holds and the AS-504 Space Vehicle was successfully launched from Launch Complex 39 at Kennedy Space Center, Florida, on Monday, 3 March 1969. Recovery of the flight crew and Command Module was successfully accomplished on 13 March 1969, for a flight duration of 241 hours 53 seconds.

Initial review of test data indicates that overall performance of the launch vehicle, spacecraft, and flight crew together with ground support and control facilities and personnel was satisfactory, and that all primary mission objectives were accomplished.

NASA OMSF PRIMARY MISSION OBJECTIVES FOR APOLLO 9

PRIMARY OBJECTIVES

Demonstrate crew/space vehicle/mission support facilities performance during a manned Saturn V mission with CSM and LM.

Demonstrate LM/crew performance.

Demonstrate performance of nominal and selected backup Lunar Orbit Rendezvous (LOR) mission activities, including:

- Transposition, docking, LM withdrawal
- Intervehicular crew transfer
- Extravehicular capability
- SPS and DPS burns
- LM active rendezvous and docking

CSM/LM consumables assessment.

Sam C. Phillips
Lt. General, USAF/ Apollo Program Director
Date: 14 Feb 1969

George E. Mueller
Associate Administrator for Manned Space Flight
Date: 17 Feb 1969

RESULTS OF APOLLO 9 MISSION

Based upon a review of the assessed performance of Apollo 9, launched 3 March 1969 and completed 13 March 1969, this mission is adjudged a success in accordance with the objectives stated above.

Sam C. Phillips
Lt. General, USAF
Apollo Program Director
Date: 30 April 1969

George E. Mueller
Associate Administrator for Manned Space Flight
Date: May 5 1969

COUNTDOWN

The terminal countdown for Apollo 9 began at T-28 hours at 10:00 p.m. EST, 1 March 1969. The only holds encountered were two planned holds: one at T-16 hours for 3 hours, and one at T-9 hours for 6 hours. The count was resumed for the last time at 2:00 a.m. EST, 3 March 1969, and proceeded to launch at 11:00:00 a.m. EST.

FLIGHT SUMMARY

The Apollo 9 mission was launched from Kennedy Space Center, Florida, at 11:00:00 a.m. EST, 3 March 1969. All launch vehicle stages performed satisfactorily, but burned slightly longer than planned, inserting the S-IVB/spacecraft combination into a nominal orbit of 102.3 by 103.9 nautical miles (NM).

After post-insertion checkout was completed, the Command/Service Module (CSM) was separated from the S-IVB, transposed, and docked with the Lunar Module (LM). The docked spacecraft was separated from the S-IVB at 4:08:05 GET (Ground Elapsed Time). After separation, two unmanned S-IVB burns were performed to place the S-IVB/ Instrument Unit on an earth-escape trajectory. After the third burn, the planned propellant dumps could not be performed.

After spacecraft separation from the launch vehicle, four Service Propulsion System (SPS) firings were made with the CSM/LM docked.

At approximately 43.5 hours GET, the Lunar Module Pilot (LMP) and the Commander (CDR) transferred to the LM. The first manned firing of the LM Descent Propulsion System (DPS) was initiated about 6 hours later. The two crewmen then returned to the Command Module (CM) for the fifth SPS firing.

At approximately 70 hours GET, the LMP and CDR again transferred to the LM for the LMP's 37-minute extravehicular activity (EVA). During this period, the Command Module Pilot (CMP) opened the CM hatch and retrieved thermal samples from the CSM exterior.

At about 89 hours GET, the CDR and LMP returned to the LM for the third time to perform the CSM/LM rendezvous. The LM primary guidance system was used to conduct the rendezvous with backup calculations being made by the CM computer. The phasing and insertion maneuvers were performed using the DPS to set up the rendezvous. The Ascent and Descent Stages were separated, followed by a concentric sequence initiation maneuver using the LM Reaction Control System. The LM Ascent Propulsion System (APS) was fired to establish the constant delta height. The terminal phase of the rendezvous began on time, and the spacecraft were again docked at about 99 hours GET. The Ascent Stage was jettisoned about 2.5 hours later. Shortly after, the APS was fired to propellant depletion. The firing lasted 350 seconds and resulted in an orbit of 3747 by 124.5 NM.

The sixth SPS firing, to lower apogee, was delayed because the +X translation to precede the maneuver was not programmed properly. However, the maneuver was rescheduled and successfully completed in the next revolution.

During the last three days, a seventh SPS firing was made to raise the apogee, and the SO65 Multispectral Photography Experiment and landmark tracking were accomplished.

Unfavorable weather in the planned landing area caused the deorbit maneuver (SPS 8) to be delayed for one revolution. This decision was made the day before splashdown and recovery forces were redeployed. Final parachute descent and splashdown were within sight of the prime recovery ship in the Atlantic Ocean. Splashdown was near the target point of 23 degrees 15 minutes north latitude, 68 degrees west longitude, as determined from the onboard computer solution. The crew were safely aboard the prime recovery ship, USS Guadalcanal, within 1 hour of splashdown. Table 1 presents a summary of mission events.

TABLE I SUMMARY OF MISSION EVENTS

EVENT	TIME (GET) HR: MIN: SEC PLANNED*	ACTUAL
First Motion	00:00:00	00:00:00
Maximum Dynamic Pressure	00:01:21	00:01:26
S-IC Center Engine Cutoff	00:02:14	00:02:14
S-IC Outboard Engine Cutoff	00:02:40	00:02:43
S-IC/S-II Separation	00:02:40	00:02:44
S-II Ignition	00:02:42	00:02:44
Jettison S-II Aft Interstage	00:03:10	00:03:14
Jettison Launch Escape Tower	00:03:16	00:03:18
S-II Engine Cutoff Command	00:08:51	00:08:56
S-II/S-IVB Separation	00:08:52	00:08:57
S-IVB Engine Ignition	00:08:55	00:09:01
S-IVB Engine Cutoff	00:10:49	00:11.05
Parking Orbit Insertion	00:10:59	00:11:15
Separation and Docking Maneuver Initiation	02:33:49	02:41:16
Spacecraft Docking	03:05:00 (Approx)	03:01:59
Spacecraft Final Separation	04:08:57	04:08:06
S-IVB Restart Preparation	04:36:12	04:36:17
S-IVB Re-ignition (2nd Burn)	04:45:50	04:45:56
S-IVB Second Cutoff Signal	04:46:52	04:46:58
S-IVB Restart Preparations	05:59:35	05:59:41
SPS Burn 1	06:01:40	05:59:01
S-IVB Re-ignition (3rd Burn)	06:07:13	06:07:19
S-IVB Third Cutoff Signal	06:11:14	06:11:21
Start LOX Dump	06:12:44	Not Accomplished
LOX Dump Cutoff	06:23:54	Not Accomplished
Start LH$_2$ Dump	06:24:04	Not Accomplished
LH$_2$ Dump Cutoff	06:42:19	Not Accomplished
SPS Burn 2	22:12:00	22:12:04
SPS Burn 3	25:18:30	25:17:39
SPS Burn 4	28:28:00	28:24:41
Docked DPS Burn	49:42:00	49:41:35
SPS Burn 5	54:25:19	54:26:12
Undocking	92:39:00	92:39:36
CSM/LM Separation	93:07:40	93:02:54
DPS Phasing	93:51:34	93:47:35
DPS Insertion	95.43:22	95:39:08
Concentric Sequence Initiation - LM RCS Burn	96:21:00	96:16:07
Constant Delta Height - APS Burn	97:05:27	96:58:15
Terminal Phase Initiation	98:00:10	97:57:59
CSM/LM Docking	99:13:00 (Approx)	99:02:26
APS Burn to Propellant Depletion	101:58:00	101:53:15
SPS Burn 6	121:58:48	123:25:07
SPS Burn 7	169:47:54	169:39:00
SPS Burn 8 (Deorbit)		240:31:15
Entry Interface (400,000 ft)		240:44:10
Drogue Chute Deployment (25,000' Approx)		240:55:08
Splashdown		241:00:54

*LV events based on MSFC LV Operational Trajectory, dated 31 January 1969. SC events based on MSC SC Operational Trajectory, Revision 2, 20 February 1969.

Premission planned deorbit was changed to permit shift in landing point due to weather and sea conditions in initial planned recovery area. One additional orbit was added.

MISSION PERFORMANCE

The significant portions of the Apollo 9 mission are discussed herein. Space vehicle systems and mission support performance are covered in succeeding sections.

TRAJECTORY

The CSM/LM/IU/S-IVB combination was inserted into earth orbit at 00:11:15 GET after a normal launch phase. The resulting orbital elements and maneuver parameters are given in Table II for all engine firings.

Four SPS maneuvers were performed prior to the first docked DPS firing. Each of the first three SPS maneuvers was made without requiring a +X translation to settle propellants. The fourth SPS maneuver was preceded by an 18-second +X translation made with the Service Module Reaction Control System (SM RCS).

The fifth docked SPS maneuver resulted in the perigee being approximately 5 nm less than planned causing the rendezvous to be initiated 4 minutes earlier. Small cutoff errors of this magnitude were expected, and real-time trajectory planning for both rendezvous and deorbit was conducted to accommodate minor adjustments in the initiation times and velocity increments. Out-of-plane components were added during the flight to certain preplanned maneuvers to provide substantial reduction in spacecraft weight without significantly changing the orbital parameters for subsequent maneuvers.

The trajectory aspects of the rendezvous exercise will be discussed in the rendezvous section.

After the Ascent Stage jettison, a separation maneuver of 3 feet per second (fps) was performed by the SM RCS. The APS engine was then fired to propellant depletion.

The sixth SPS maneuver was delayed one revolution when the accompanying ullage burn did not occur at the proper time, but was completed nominally.

The seventh SPS maneuver was restructured in real time to provide a desired higher burn time and was successfully accomplished.

The deorbit maneuver was made over Hawaii during revolution 152, and CM/SM separation was performed. The CM landed at 241:00:53 GET near 23 degrees 15 minutes north latitude and 68 degrees west longitude.

TABLE II SUMMARY OF MANEUVERS

	BURN TIME (SEC)			Delta V (FPS)			RESULTANT ORBIT		
	*Prelaunch PLANNED	Real Time PLANNED	ACTUAL	*Prelaunch PLANNED	Real Time PLANNED	ACTUAL	*Prelaunch PLANNED	Real Time PLANNED	ACTUAL
First Service Propulsion	5.0	4.96	5.2	36.8	36.8	36.6	125.2 X 108.7	128.2 X 110.2	127.6 X 111.3
Second Service Propulsion	111.3	111.2	110.3	949.6	850.6	850.5	190.2 X 109.1	189.8 X 107.7	192.5 X 110.7
Third Service Propulsion	280.0	281.9	279.9	2548.2	2570.7	2567.9	268.2 X 1111.3	270.3 X 109.4	274.9 X 112.6
Fourth Service Propulsion	28.1	28.4	27.9	299.4	300.9	300.5	268.7 X 111.4	273.8 X 109.3	275.0 X 112.4

	BURN TIME (SEC)			Delta V (FPS)			RESULTANT ORBIT		
	*Prelaunch PLANNED	Real Time PLANNED	ACTUAL	*Prelaunch PLANNED	Real Time PLANNED	ACTUAL	*Prelaunch PLANNED	Real Time PLANNED	ACTUAL
First Descent Propulsion	367.0	370.6	372.0	1734.0	1744.0	1737.5	267.6 X 111.8	269.9 X 109.1	274.6 X 112.1
Fifth Service Propulsion	41.5	43.2	43.3	552.3	575.4	572.5	130.2 X 129.7	129.8 X 129.8	131.0 X 125.9
Ascent Propulsion Firing to Depletion	389.0**	444.9**	362.4	6074.9**	7427.5**	5373.4	4673.3 X 128.9**	6932.3 X 125.9	3760.9 X 126.6
Sixth Service Propulsion	2.4	1.33	1.40	62.7	38.8	33.7	127.9 X 94.6	120.2 X 104.8	123.1 X 108.5
Seventh Service Propulsion	9.9	25.0	24.9	252.8	653.3	650.1	238.7 X 93.9	250.4 X 97.9	253.2 X 100.7
Eighth Service Propulsion	11.7	11.6	11.7	323.3	325.0	322.7	241.8 X -15.1	238.5 X ---	240.0 X -4.7

NOTES: Prelaunch Planned refers to Apollo 9 Spacecraft Operational Trajectory, Revision 2, 20 February 1969. APS burn to depletion planned for unattainable apogee value to insure propellant depletion cutoff.

EXTRAVEHICULAR ACTIVITY

Extravehicular activity (EVA), planned for the third day, was reduced from 2 hours 15 minutes to about 1 hour of depressurized LM activity. This change was made because the LMP experienced a minor in-flight illness during the first two days of the mission.

Preparation for EVA began at approximately 71 hours GET. The CDR and the LMP were in the LM and the CMP in the CM. At approximately 73 hours GET, after donning the Portable Life Support System (PLSS) and the Oxygen Purge System (OPS), the LMP egressed through the forward hatch and moved to the external foot restraints on the platform. During this time the CM was depressurized and the side hatch was opened. Thermal sample retrieval was photographically recorded with the sequence cameras. The LMP used the handrails to evaluate body control and transfer techniques. Ingress was completed at about 74 hours GET. Both hatches were then secured and the vehicles repressurized. The PLSS was successfully recharged with oxygen and water. The lithium hydroxide cartridge from the system was returned to the CM for post-flight metabolic analysis. The repressurization cycles for both vehicles were nominal, and post-EVA procedures were followed without difficulty.

RENDEZVOUS

The CDR and the LMP transferred to the LM on the fifth day for the rendezvous. The rendezvous exercise began on schedule with a 5-fps separation maneuver using the SM RCS.

A phasing maneuver of 90.5 fps was performed with the LM DPS about 2.8 nm from the CSM. Approximately 12 nm above and 27 nm behind the CSM, the DIPS was used to impart a 43.1-fps insertion velocity to the LM. At a range of 75 nm from the CSM, the Ascent and Descent Stages of the LM were separated, and a concentric sequence initiation maneuver of 40.0 fps was made with the LM RCS.

Approximately 10 nm below and 78 nm behind the CSM, the constant delta height maneuver was performed with the APS imparting a velocity change of 41.5 fps. The terminal phase began on time with a 22.3-fps LM RCS maneuver.

Braking maneuvers were conducted on schedule, and stationkeeping was maintained at a distance of approximately 100 feet so that photographs could be taken from both vehicles. Docking was successfully completed at about 99 hours GET. Problems were experienced in using the Crewman Optical Alignment Sight (COAS) in both vehicles during docking. The combination of a bright CM, a dimly lighted CM target, and a relatively dim reticle in the alignment sight made LM docking a difficult task.

LM rendezvous navigation and maneuver targeting using both the primary and the backup guidance systems were satisfactory. Radar data were successfully used, both automatically by the primary system and through manual insertion in the Abort Guidance System, to correct rendezvous state vectors. Maneuver solutions from both onboard systems and from ground computations appeared to correlate closely. The crew selected the primary system solutions for all maneuvers through the first midcourse correction performed after terminal phase initiation.

Rendezvous navigation and mirror-image targeting in the CM were performed satisfactorily; however, loss of the LM tracking light prevented sextant measurements from the CM when both vehicles were in darkness. Preliminary data indicate that CM maneuver calculations for terminal phase initiation were satisfactory.

FLIGHT CREW PERFORMANCE

Crew performance was excellent throughout the mission, and the flight was conducted essentially in accordance with the nominal plan.

Preparation for transfer to the LM required longer than anticipated, primarily because of the time required for the crewmen to don the space suits. The suit supply hoses were a source of interference and also contributed to the longer preparation time. As a result, about 1 hour was added to the preparation time for subsequent transfer.

Visual and photographic inspection of the entire spacecraft was accomplished after rendezvous and before docking.

FLIGHT CREW BIOMEDICAL EVALUATION

The launch was postponed for 72 hours because of symptoms of upper respiratory infections in all three crewmen. Physical examinations 3 hours before launch revealed no infection.

The planned medical operations were conducted as scheduled except that the LMP experienced some nausea and vomiting prior to and following the initial transfer to the LM.

Plans for EVA were modified because of the LMP's illness. The physiological parameters were essentially normal throughout the mission. The LMP's work rate during EVA was on the order of 500 Btu/hr.

FLIGHT CONTROL

Flight control performance was satisfactory in providing operational support for the Apollo 9 mission. Minor spacecraft problems were encountered, but none was such that either the mission operations or the flight plan was significantly altered.

Early in the mission, a caution and warning light on Hydrogen Tank 1 was observed just prior to an automatic cycle of the heaters. This condition persisted and the crew had to be disturbed during a rest period at 81 hours GET to increase the hydrogen tank pressure.

On the third day, the crew were about 1 hour behind the timeline, resulting in canceling all the planned communications tests except the LM secondary S-band test and the LM two-way relay with television.

On the fourth day, the EVA was abbreviated and the external transfer from the LM to the CM was not performed. The activity was restricted to the LM forward platform because of concern about the LMP's earlier illness and proper readiness for the rendezvous on the following day.

At approximately 78 hours GET, after the tunnel hardware had been installed, a crewman made an unplanned return to the LM to open a circuit breaker. This change shortened the rest period about 30 minutes.

On the fifth day, LM activation was performed approximately 40 minutes early to insure an on-time rendezvous initiation.

The LM VHF telemetry and S-band power amplifier were lost for 6 and 12 hours, respectively, after the APS firing to depletion. These failures were expected because of the lack of cooling. The electrical system capability for this spacecraft was several hours longer than predicted. LM support terminated at 113:42:00 GET.

On the sixth day, the sixth SPS maneuver was delayed by one revolution. The crew reported that the +X translation did not occur. A procedural error was made in loading the CM computer, since the proper SM RCS quads were not selected. The computer was reloaded, and one revolution later, the maneuver was made satisfactorily.

On the eighth day, the seventh SPS maneuver was increased to 25 seconds in duration to permit a test of the Propellant Utilization and Gaging System (PUGS).

RECOVERY

Recovery of the Apollo 9 Command Module and crew was completed in the West Atlantic by the prime recovery skip, USS Guadalcanal. The following table is a list of significant recovery events on 13 March 1969:

EVENT	EST
First VHF contact	11:51 a.m.
First beacon and voice contact	11:57 a.m.
First visual contact	11:59 a.m.
Landing	12:01 p.m.
Swimmers deployed	12:07 p.m.
Flotation collar installed	12:14 p.m.
CM hatch open	12:27 p.m.
First astronaut aboard helicopter	12:39 p.m.
All astronauts in helicopter	12:46 p.m.
Astronauts on deck	12:50 p.m.
CM aboard recovery ship	2:13 p.m.

The CM remained in the stable I flotation attitude. Sea-state conditions were very moderate at the recovery site.

SYSTEMS PERFORMANCE

Engineering data reviewed to date indicate that all mission objectives were attained. Further detailed analysis of all data is continuing and appropriate refined results of systems performance will be reported in MSFC and MSC technical reports. Summaries of the significant anomalies and discrepancies are presented in Tables III, IV, and V.

TABLE III

LAUNCH VEHICLE DISCREPANCY SUMMARY

DESCRIPTION	REMARKS
Oscillations occurred in the S-II center engine chamber pressure and the S-II structure late in the burn. Oscillations have occurred on four flights and five static firings, but only after 320 seconds of S-II burn.	Apparently caused by coupling between the center engine and the stage structure. Fix will be early center engine cutoff at 299 seconds on Apollo 10.
S-IVB APS Module No. 2 helium supply pressure decayed slowly.	Leak in teflon seals upstream of the regulator. Change of seal material to rubber has been approved. Closed.
S-IVB helium regulator lock-up pressure exceeded the redline during countdown, and the helium pneumatic pressure was high throughout the mission.	Internal leakage in regulator caused by wear on poppet. Modified regulator has been tested and installed on S-IVB-505. Redline has been raised from 585 to 630 psi.
S-IVB third burn anomaly: Gas generator pressure spike at start, engine chamber pressure oscillations, loss of engine control pneumatic pressure, abnormal attitude control system oscillations, decrease in engine performance during burn, and inability to dump residual propellants after burn.	Caused by extreme out-of-spec engine start conditions which resulted in excessive engine chamber pressure oscillations and possible gas generator damage, followed by loss of pneumatic system. There is no evidence that the causes of this anomaly are applicable to an in-spec engine start. The flight mission rules allowing restart with recirculation systems inoperative are being revised for Apollo 10.

TABLE IV

COMMAND/SERVICE MODULE DISCREPANCY SUMMARY

DESCRIPTION	REMARKS
Unable to translate the CSM to the left. Propellant isolation valves in two SM RCS quads were found to be closed.	Apparently caused by mechanical shock at CSM/S-IVB separation. The crew will check the valve positions after separation on Apollo 10 and subsequent missions.
Master alarm occurred coincident with hard docking without any accompanying annunciator.	Caused by a sensor transient or a momentary short circuit due to mechanical shock. Also occurred during the CSM 106 docking test.
During the third SPS burn, eight master alarms occurred because of indications of propellant unbalance.	Caused by erroneous readings from the primary probe in the SPS oxidizer tank. The master alarm and warning functions from the PUGS have been deleted on CSM 106 and subsequent spacecraft. Closed.
The scanning telescope mechanism jammed frequently when driven manually, but worked normally in automatic mode.	A pin from a counter drum was found wedged in a split gear. Units on Apollo 10 and subs will be replaced with units that have been inspected. Closed.
Fuel Cell No. 2 condenser outlet temperature exceeded the normal range several times.	The bypass valve that controls coolant temperature operated improperly because of contamination in the glycol. For subsequent missions, Block 1 valves which are less susceptible to contaminants will be installed and the radiators will be vibrated and flushed 30 to 45 days before launch. Closed.
Automatic control of the pressure in the cryogenic hydrogen tanks was lost and pressure was controlled manually.	Probably caused by an intermittent open circuit in the motor switch control circiut. No hardware change will be made. Closed.
The first two attempts to undock were unsuccessful because the release switch was not held long enough. Before the 2nd docking, the "flag" check showed the capture latches on the probe were not cocked; recycling the switch produced a cocked indication.	The Apollo Operations Handbook has been revised to clarify the procedure for extending the probe. Closed.